100 THINGS
TO SEE
IN THE
SOUTHERN
NIGHT
SKY

100 THINGS TO SEE IN THE SOUTHERN NIGHT SKY

From Planets and Satellites to Meteors and Constellations, Your Guide to Stargazing

DEAN REGAS

Adams Media

New York   London   Toronto   Sydney   New Delhi

Adams Media
An Imprint of Simon & Schuster, Inc.
57 Littlefield Street
Avon, Massachusetts 02322

First Adams Media trade paperback edition June 2018

ADAMS MEDIA and colophon are trademarks of Simon & Schuster.

For information about special discounts for bulk purchases, please contact Simon & Schuster Special Sales at 1-866-506-1949 or business@simonandschuster.com.

The Simon & Schuster Speakers Bureau can bring authors to your live event. For more information or to book an event contact the Simon & Schuster Speakers Bureau at 1-866-248-3049 or visit our website at www.simonspeakers.com.

Interior design by Michelle Kelly
Interior images by Eric Andrews

Manufactured in China

10 9 8 7 6 5 4 3 2 1

Library of Congress Cataloging-in-Publication Data
Regas, Dean, author.
100 things to see in the southern night sky / Dean Regas.
Avon, Massachusetts: Adams Media, 2018.
Series: 100 things to see.
Includes index.
LCCN 2018003624 (print) |
LCCN 2018006728 (ebook) | ISBN 9781507207802 (pb) | ISBN 9781507207819 (ebook)
LCSH: Astronomy--Amateurs' manuals. | Constellations--Amateurs' manuals. | Southern sky (Astronomy)--Observers' manuals. | Southern sky (Astronomy)--Amateurs' manuals.
LCC QB63 (ebook) | LCC QB63 .R3625 2018 (print) | DDC 523--dc23
LC record available at https://lccn.loc.gov/2018003624

ISBN 978-1-5072-0780-2
ISBN 978-1-5072-0781-9 (ebook)

Contains material adapted from the following title published by Adams Media, an Imprint of Simon & Schuster, Inc.: *100 Things to See in the Night Sky* copyright © 2017, by Dean Regas, ISBN 978-1-5072-0505-1.

Contents

Introduction

Constellations. Stars. Planets. Nebulas. Satellites.

The universe is full of wonders beyond our imagination. And while much of what's out there in space can only be seen by powerful telescopes or theorized about by professional astronomers, there are many things you can see in the night sky just by walking out your door and looking up.

In *100 Things to See in the Southern Night Sky* you'll find 100 entries, each telling you about particular objects in the night sky that you can find with either your naked eye, binoculars, or a backyard telescope. Each entry tells you what type of heavenly object you are looking for (planet, star, etc.), how hard it is to find, what the item looks like and its mythology, and where and when to look for it. You'll also find a variety of star maps scattered throughout the entries that will show you exactly where to look when you turn your eyes toward the stars.

The tips, techniques, and informative bits that you'll find in the book are the same ones that I used when I first started heading outside every clear night to let the sky become my classroom. They're the same tips that turned me into a starstruck astronomer. Hopefully they'll do the same for you, and help you fall in love with the heavens!

But what exactly will you find in this book? Well, first you'll learn how to safely observe the Sun, take in sunsets and sunrises, chart the seasons, and observe sunspots. Then you'll explore the phases and features of the Moon and uncover the hidden secrets of the five planets closest to Earth: elusive Mercury, dazzling Venus, ruby-red Mars, giant Jupiter, and spectacular Saturn.

Then you will delve into the heart of the book and study the art of stargazing. You will learn how to identify major stars and constellations throughout every season of the year. From beginner star patterns like Crux, the Southern Cross and Orion's Belt to more challenging constellations like Delphinus, the Dolphin and Pavo, the Peacock, you will soon be able to recognize dozens of stars in the night sky and retell their ancient mythological stories.

Finally, you'll discover the tricks to observing man-made satellites and supreme heavenly shows such as meteor showers and eclipses.

And, in case you're not sure exactly what you need to do, you'll find a chapter packed with information on ideal stargazing locations and sky conditions, what type of equipment you need (in some instances, you just need to use your two eyes and bare hands), the best time to stargaze, and more. So, whether you're looking to learn about the night sky on your own or are looking to get your kids excited about what lies in the skies above, this book is for you. I challenge you to go outside and find every one of the night sky objects detailed here. When you do that, you may be converted to the stars like me, and you may find that you're an astronomer too!

How to Use This Book

The universe is yours to behold and this book can be your beginner's guide. It will focus on 100 of the most amazing astronomical objects that you can see in the night sky. You'll find basic information and tips to locate each individual object, often with the help of accompanying charts and graphics. But some general rules and guidelines are necessary to start properly. Let's talk about viewing locations, breaking up the sky, various viewing conditions, and any equipment you may have or want to get.

Location, Location, Location

This edition of this book is intended specifically for viewing audiences in Australia, New Zealand, South Africa, and South America. It is designed to help you view the night sky from the mid-southern latitudes, which means that if you're living or traveling between 15 degrees south and 45 degrees south latitude, this is the book for you. Your perspective on the heavens does not change much when you travel east and west, but it does change when you trek north or south. There will be some useful information in this book no matter where you live, but my observing tips and star charts are mainly geared for stargazers living in the mid-southern latitudes.

Break Up the Sky

That said, no matter where you are on Earth, if you stargaze long enough you will notice that the stars, constellations, planets, and the Sun and Moon slowly move across the sky. Hour by hour, minute by minute, they shift as one body. Ancient astronomers pictured the dome of heaven circling around an unmoving Earth. It was as if the gods were manipulating a great wheel behind the scenes that made everything rise and set and circumnavigate the globe once a day.

For beginning stargazers to really experience this motion it helps to break up the sky by the four cardinal directions: south, north, east, and west. When you face south and watch the stars, you will see that they behave quite differently than when you face any other direction. Take some time to sit outside under the stars and note the positions of several bright stars all around you and place them in reference to landmarks like houses, trees, and mountains. If you live between 15 and 45 degrees south latitude, on every night of the year this is what you will notice:

- **South:** When you face south, the stars will seem to move in a clockwise circle around a seemingly empty point in the sky. Unlike in the Northern Hemisphere where stargazers can watch the stars endlessly circle a stellar landmark, the North Star, aka Polaris, there is no equivalent in the Southern Hemisphere. However, you will still see recognizable star patterns like Crux, the Southern Cross and the Southern Triangle as well as the bright stars Alpha and Beta Centauri slowly appear to revolve around a point above the southern horizon. Many of the stars nearest to this apparent pivot point never rise or set but endlessly seem to circle. Astronomers call these circumpolar stars.

- **West:** When you face west, the motion seems quite different. Stars, planets, the Moon, and the Sun appear to move diagonally down and to the left until they set below the horizon.

- **East:** When you face east, everything rises and travels diagonally up and to the left.

- **North:** When you turn north, the celestial objects move from right to left and almost crawl across the northern horizon.

Knowing how things move in each direction will help you know what stars and constellations are coming next and which objects might be setting soon.

Note: If you are in the Northern Hemisphere, your motions seem different. Objects still rise in the east and set in the west, but they travel up and to the right and down and to the right, respectively. Objects in the southern sky move left to right, and the stars in the northern sky completely circle the pole in a counterclockwise direction. No matter where you live, all of these apparent motions are caused by one thing: the daily rotation of the Earth.

Prime-Time Stargazing

You probably won't be surprised to hear that more people watch the stars during the prime-time hours (around 8 or 9 p.m.) than any other time of day. The observing suggestions throughout the book for both the stars and constellations usually reflect this timing. So, for example, when I describe a group of stars as winter constellations or summer constellations, I mean that they are best observed during the evenings of their respective seasons. That said, please note that seasonal stars can be seen in other months, but at different times of night. For example, you can see Orion in the summer evenings, but you can also find him gracing the sky on winter mornings just before dawn.

Sky Conditions

Obviously if you're going to head outside to stargaze, you'll want to pay attention to the weather conditions. The first thing you need to know is that clouds are the bane of the astronomer's existence. Avoid them at all costs. Whether you're viewing with your naked eye, telescope, or binoculars, you will need clear skies to see the maximum number of stars. Even the best telescopes cannot see through thick cloud cover.

You also need to take light pollution into consideration. As cities grow larger so do their domes of light. The more light that we shine up, the less light that the stars shine down. From urban locations you can sometimes barely make out the brightest twenty to thirty stars in the sky, called the first magnitude stars. But the farther you travel away from light pollution, the more stars you can see. In the suburbs you may see second, third, and fourth magnitude stars, which include the brightest 700 stars in the sky. In a truly dark sky you can see to the fifth and even sixth magnitude stars to behold up to 6,000 stars. So in the countryside, you can sometimes observe stars with the naked eye that are 100 times fainter than the stars people see in large cities. Additionally, from a dark sky you can see the Milky Way and even a few other deep sky objects that are between 1,000 and 2.5 million light-years away.

Most of the objects described in this book are first or second magnitude objects and are easily visible from light-polluted or semi-light-polluted skies. Throughout the book, you'll find info on how easy (or not-so-easy) it is to see these night sky objects. The first and second magnitude objects are generally labeled as Easy; third and fourth magnitude objects are labeled as Moderate; and anything that's harder to see than that is labeled Difficult. Such Difficult objects include the Beehive Cluster and the Andromeda Galaxy. Hopefully these designations will spur you to seek out the darkest locations so that you can see a sky full of stars.

The Moon itself can actually negatively affect your stargazing. While a big, beautiful Full Moon may add to the romance of an evening, it also puts out a lot of light. This added moonlight can wash out the fainter light of the stars and limit the number of objects you can detect. For the optimal stargazing experience, plan your observations around the New Moon and crescent phases.

Equipment

When you are trying to find faint objects in the sky with the naked eye, it may take you a while to see them. Even if you are away from light pollution and clouds, there is no Moon in the sky, and you have ideal viewing conditions, you still might have trouble finding some of these naked-eye objects. Do not be deterred. The objects are still there and can usually be seen without any additional equipment. However, if you're struggling to find them, then using a pair of binoculars or peering through a telescope can help you enhance the experience.

You see, the more light you can gather, the more you can magnify an image and the more details you can see. And that is precisely what binoculars and telescopes do. With the exception of viewing the Sun, if you have binoculars or a telescope try pointing them at every object in this book for easier viewing. (Note: in this book I will tell you how to safely view the Sun.)

Use Your Hands to Find Your Angles

Amateur astronomers break up the sky into degrees and use them to determine how high an object is in the sky or the angular separation between two objects (how far apart they seem in the sky). In order to help you locate objects, you'll find references to degrees in the entries throughout the book. Conveniently, you have the built-in tools you need to measure angles yourself. Simply picture the sky as a dome or hemisphere. This gives you 180 degrees of sky to work with—from horizon to horizon. If an object is straight overhead (the zenith), it is 90 degrees above the horizon. A star halfway up the sky is therefore at 45 degrees. A star that is one-third of the way above the ground is at about 30 degrees.

To make finer angular measurements, you can use your hand at arm's length. If you extend your arm and make a fist, that fist will cover approximately 10 degrees of the sky. If you open your hand and spread out your fingers, the space between the end of your thumb and the tip of your little finger will measure about 25 degrees. The width of your little finger at arm's length is only 1–2 degrees. These measurements are not perfect, but making rough estimates like these can help you hop from a constellation you know to one you want to find. This method can also help you locate fainter stars that may be near brighter ones.

MEASURING ANGLES

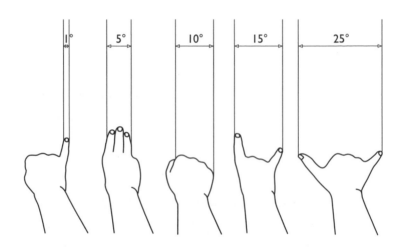

Patience and Practice

The best teacher of astronomy is the sky itself. Get outside every clear night and watch and learn. If you do this regularly, after less than one month the sky above will all start to make sense. You'll not only come to know the major stars and constellations, the planets, Moon, and Sun but will develop an appreciation for the heavens above. Let's get started!

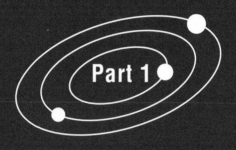

THE SUN, MOON, AND NAKED-EYE PLANETS

How do you get started studying the universe? I recommend doing what the ancients did: observe the brightest things first. In this section of the book we will explore the brightest, most noticeable objects in the daytime and nighttime sky.

We will start off with the Sun. I know it may seem funny for a book titled *100 Things to See in the Southern Night Sky* to begin by talking about the brightest thing in the daytime sky, but there are so many cool aspects of the Sun that we should check out more closely. Observing the daily motion of the Sun taught our ancestors about time. Monitoring its changing path from month to month taught them the seasons. And now modern astronomers can peer deeply through the sunlight and tell us how hot the sun is, how far away it is, and what it is made of.

You'll also learn about the Moon. This is the second brightest object in the sky and it presented a more challenging cycle for ancient astronomers to decipher. I'll show you how to make sense of its phases, when and where to look for it, and how best to see it in a telescope.

Then we will fly to the five planets that are visible to the naked eye: Mercury, Venus, Mars, Jupiter, and Saturn. Venus shines in as the third brightest celestial object with Jupiter and Mars coming in at numbers four and five, respectively. Although there are several stars that can be brighter than Mercury and Saturn, these planets make such interesting sights and appear much different than stars around them.

The objects detailed in this section—the Sun, the Moon, and the five naked-eye planets—are all so bright and so stunning that they are difficult to miss. Let's check them out!

THE SUN

What Is It?
Our closest star

Difficulty Level
Easy

Description
The Sun is our nearest star (about 150 million kilometers from Earth) and is the major source of light and heat for all the planets, moons, and asteroids in the solar system. At roughly 1,400,000 kilometers wide, the Sun is about 1,300,000 times larger than Earth. It is so massive that it holds all of the planets in steady orbits.

Sun Safety
Whenever you view the Sun, think before you look. Do not look at the Sun for more than a moment unless you are 100 percent sure you are using safe and proven viewing methods. Never use common homemade filters including CDs, smoked glass, exposed film, X-ray transparencies, foil wrappers, or Mylar balloons. Your vision can be permanently damaged with even short periods of direct exposure to sunlight.

The best way to directly observe the Sun safely with the naked eye is by looking through either a sheet of #14 welder's glass or specially made eclipse shades. You can get #14 welder's glass at welding supply stores, and several companies sell effective eclipse shades online. These inexpensive items will allow you to observe the Sun's disc, partial eclipses, and maybe even a large sunspot. If you use the welder's glass, be sure to use only #14. That is the only shade dark enough to protect your eyesight—anything less is unsafe.

You'll notice that the Sun seems surprisingly small in the sky when you examine it by looking through the glass or the glasses. The Sun is humongous, but after you block out most of the surrounding glare, it looks like a tiny circle of light. After all, you are looking at the Sun from 150 million kilometers away! It's not exactly close to you.

You can also take pictures of the Sun through a sheet of welder's glass or through eclipse glasses. Place the welder's glass or eclipse glasses in front of a camera, aim it at the Sun, and then review the image on your screen. Center it, adjust the contrast, and snap away. You'll have pictures of the Sun! Welder's glass or eclipse glasses act like a filter that will reduce the glare and will protect your camera in the same way it protects your eyes.

See Features on the Sun

Now that you have your safe solar-viewing glasses, what can you see? To the naked eye, on a normal day, you can't make out a lot of detail on the Sun. But sometimes you can pick out larger sunspots or, on very rare occasions, a solar eclipse.

Sunspots are darker, cooler regions on the Sun where magnetic disturbances cause stellar material to erupt from the surface. Through safe glasses, they look like tiny black blemishes on the glowing disc of the Sun. Only large sunspots can be detected with the naked eye, and finding them can be a test of your eyesight. They are more frequent during solar maximum, which is a varied period of time that occurs about every eleven years, when solar activity ramps up. But sunspots can pop up at any time, so it is always worth taking a look. If you are lucky enough to see a sunspot with the naked eye, remember that the spot is larger than Earth!

The best time to observe the Sun is during a solar eclipse. The most common eclipse is a partial solar eclipse, which is when the Moon covers only a portion of the solar disc. It looks like someone took a bite out of the Sun. Although solar eclipses occur about twice a year, they are localized events, which means each eclipse is visible only from certain areas on the globe. For any one location, you may see a solar eclipse about once every three to six years.

The best upcoming solar eclipses that will be visible from locations in South America will occur on July 2, 2019, and December 14, 2020; in Australia on April 20, 2023; and in Australia and New Zealand on July 22, 2028. Use your eclipse shades to get a safe view of these cool alignments of Sun, Moon, and Earth and admire the fact that astronomers, after watching the heavens for centuries, can now predict them well in advance!

See Several Sunsets and Sunrises

Even though you experience one sunrise and one sunset every day, you may not have considered the complex series of steps that happen during these events. To start, you want to look for several subtle changes around the sky in addition to the Sun. For a sunrise, simply get up while it is still dark and then watch as the sky slowly brightens. While the Sun is almost ready to rise in the east, look behind you to the west. You will see a slightly darker piece of sky just above the horizon. That is Earth's shadow. When you look back to the east again you will notice every layer of the atmosphere igniting in different warm tones. The rising but still unseen Sun turns the atmosphere closest to the horizon from deep blue to ruby red in a matter of minutes. The wind may change as the temperature rises in anticipation of the rising Sun. And then, if you have a clear view to the horizon, you will see the top of the Sun peek up above it. It will take a few minutes for the Sun to fully clear itself from the horizon, but when it does, the day has broken.

Experiencing a sunset is just as powerful (and you don't have to get up extra early). The play of light and darkness is reversed as you face west to see the Sun slowly set. The sky gradually turns from light to dark through every color in the rainbow. Just before sunset, turn around and face east to see the shadow of the Earth cast on the lower reaches of the atmosphere. Then face west again and bask in the final rays of sunlight as the Sun dips below the distant horizon. Night has fallen.

Sense the Seasons

You may have heard that "the Sun rises in the east and sets in the west." Well, that's mostly accurate. If you watch several sunrises or sunsets over the course of a few months, you will notice that the Sun does not always rise or set in the same place. And you may have seen that the Sun is much higher in the sky during the summer months than it is in the winter. Observing these changes in the position of the

Sun is one of the oldest astronomical practices and can be done very easily with the naked eye.

In the following diagram you can see how the Sun moves across the sky in different seasons from any location in the mid-southern latitudes. Around June 21, the winter solstice (the bottom line on the diagram), the Sun rises north of east, cuts low across the northern sky, and then sets north of west. Three months later, on the spring equinox (the middle line of the diagram), the Sun rises due east, reaches a higher altitude in the north, and sets due west. Notice that the lengths of the Sun's arcs are different—in spring the Sun rides higher in the sky and stays visible for a much longer period of time than it does in the winter. Around December 21, the summer solstice (the top line of the diagram), the Sun rises south of east, travels extremely high in the northern sky, and then sets south of west. This is called the longest day because you experience the most minutes of daylight and the least amount of darkness for the year. The path of the Sun on the Autumnal Equinox is almost exactly equal to the arc it cut on the Spring Equinox.

Sketch what you see along the horizon; note and date the changes throughout the year. You will quickly learn that we need to modify the rule to instead read, "The Sun rises in the east-ish and sets in the west-ish."

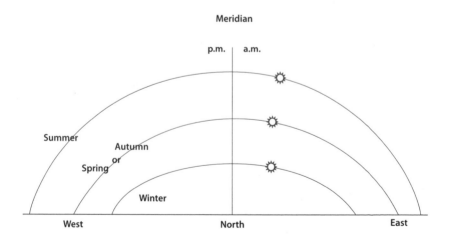

THE MOON

What Is It?

Earth's natural satellite

Difficulty Level

Easy

Description

For thousands of years our ancestors have gazed up at the silvery glow of the Moon and tried to make sense of its rhythm and aura. What are the dark and light portions of the Moon? And why does the Moon change shape and size? Memorialized in literature, song, and dance since the dawn of time, the Moon tugs at our romantic nature and inspires our soul.

The Moon is our closest neighbor in space: a ball of rock nearly 3,500 kilometers in diameter that circles Earth from a distance of about 386,000 kilometers. It is the easiest and most dynamic object in the sky to view with the naked eye. Not only can you observe the Moon in the night sky during parts of each month, you can often catch sight of it during the daytime. It changes positions from night to night and rises and sets in different places regularly. And for a few days each month, you can't see it at all.

Observe the Phases of the Moon

You've noticed that the Moon doesn't look the same every time you look up into the night sky. Each night, it goes through a different phase. The images in the following chart illustrate these phases as seen from the Southern Hemisphere.

So what causes all these shapes? The answer is light! The Moon doesn't generate any light of its own—it shines with reflected sunlight. Light travels 150 million kilometers from the Sun, bounces off the surface of the Moon, and then travels another 386,000 kilometers to reach your eyes. The phase you see depends on where the Moon is in its orbit around Earth. When the Moon is between the Sun and Earth, the only half of the Moon that is lit faces the Sun. This is the New Moon phase, and from Earth the Moon appears completely dark. When the Sun, Moon, and Earth form a right angle, you will see only half of the Moon lit (that happens at first and last quarter). And when the Moon is on the opposite side of Earth with respect to the Sun, you will see the entire Moon illuminated. That's a Full Moon.

MOON PHASES

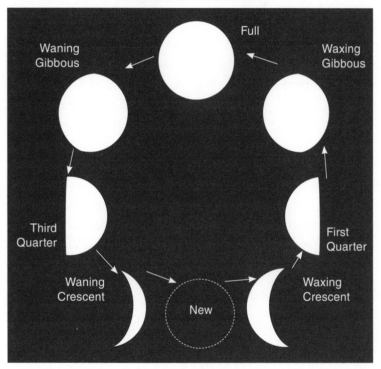

It takes the Moon about 29.5 days to go through its entire cycle of phases, and so every night the Moon's phase changes slightly as different parts of the lunar surface are lit by the Sun, while others plunge into shadow.

Some people incorrectly believe that the phases of the Moon are caused by Earth casting a shadow on it. That makes sense when you see a crescent Moon, but how can a round Earth create a straight-line shadow like we see with a First Quarter Moon, or the curve on a gibbous moon? Earth can't make those phases of the Moon that you can see all month.

Observe the Moon every clear night during a month and check out not only how the phase changes but how its location in the sky and time that it is visible changes. Remember, to complete its orbit, each night the Moon moves about one twenty-ninth the way around Earth. So it will appear to shift farther to the east 12–14 degrees from night to night. Although that's just a little larger than the width of your fist at arm's length, the shift is fast (astronomically speaking). This motion of the Moon around Earth also causes the Moon to rise about 50 minutes later each night. That time varies depending on the season and the phase of the Moon, but test it out. Make a Moon journal and sketch the position of the Moon every day at the same time. Soon you may be able to predict where and when the Moon will travel in the future.

Drink in Some Earthshine

Have you ever looked at a crescent Moon and noticed that the dark part of the Moon looks like a faintly glowing, gray ball? This effect is caused by earthshine, where the light of the Sun shines on the Earth and reflects up onto the Moon. Earthshine brightens the Moon enough so that you can see all of it against the darkness of the sky. Look for this effect during either a Waxing Crescent Moon (visible just after sunset) or a Waning Crescent Moon (visible just before sunrise).

Experience the Moon Illusion

Have you noticed that the Moon looks bigger when it is near the horizon? Well, although the Moon appears larger near the horizon, it really isn't. This effect is called the Moon illusion and scientists have debated about it for years. Astronomers first thought that the Moon appeared larger on the horizon because we have reference points like trees and buildings. But now they believe that the Moon illusion is caused by how we perceive the sky. We picture space on the horizon to be farther from us than overhead. Since the Moon is actually the same size, the "farther" Moon becomes larger in our minds. The illusion takes place in our brains. You can combat the Moon illusion by looking at it upside down, between your legs. For some reason the Moon will look normal-sized. Try it. It really works.

See a Supermoon!

The Moon's orbit around Earth is not a perfect circle. It travels in a slightly stretched oval shape, called an ellipse. This elliptical path means the Moon changes its distance from the Earth. When the Moon is slightly closer to Earth, it can appear slightly larger in the sky. The closest Full Moon in a calendar year is known as a Supermoon. A naked-eye observer can't differentiate the Moon's apparent size from night to night. However, when one compares a Supermoon to the farthest Full Moon, a so-called Puny Moon, the variance is dramatic. The Supermoon is more than 50,000 kilometers closer to Earth, and consequently it appears 14 percent larger in diameter with a 30 percent larger surface area than the Puny Moon.

In the second century B.C. the Greek astronomer Hipparchus noticed this changing Moon size with his naked eye. He constructed a device called a diopter, a two-meter-long stick with a sighting circle on the far end that could help him measure very small angles in the sky. After utilizing the diopter through several lunar cycles, Hipparchus discovered that the Moon's angular size (how large it appears to be in the sky) changed. Any time you observe a Full Moon, see if you can tell the difference between a Supermoon and a Puny Moon.

Sailing the Seas and Scaling the Highlands on the Moon

There are two main terrains on the Moon that you can see with the naked eye: the brighter areas are called the highlands, and the darker areas are called the seas, or maria. The highlands are older structures and include many mountain peaks and a large number of craters. But your eye will also be drawn to the darker seas that cover 30 percent of the lunar surface.

When you observe a Full Moon, notice that the seas are mostly circular in shape. Each sea is really the remnant of a gigantic crater. Eons ago, the Moon was hit so hard by comets and meteors that it literally cracked. Over time, magma seeped out of these cracks and filled in the craters to form the seas. Some of these seas bear features that have given rise to their peaceful names, like the Sea of Serenity and the Sea of Nectar. Others have more ominous titles, such as the Sea of Crises and the Sea of Rains. The most famous lunar sea is the Sea of Tranquility. American astronauts Neil Armstrong and Buzz Aldrin planted the first human footprints there on July 20, 1969.

The outlines of the highlands and the seas can often spark your imagination, and if you look long enough you may start to see shapes and patterns in these splotches on the face of the Moon. At certain Moon phases the play of light and shadow on the lunar surface may inspire you to see the "Man in the Moon." This configuration of seas and highlands gives many people the impression of seeing a human face. With a little more imagination you can connect the seas in such a way as to form a rabbit with two long ears and a fluffy tail. Or maybe it will look like a man carrying a bucket of water or perhaps a lady wearing a necklace. When you gaze at the Moon, let your imagination run wild.

Observing the Moon with Binoculars and Telescopes

With a little magnification you can start to see more details and features on the Moon. Even a simple pair of binoculars can provide an amazing view of the lunar surface. Most binoculars can be attached to a tripod with the aid of a simple adapter. If you can mount your binoculars on a steady tripod, you can watch the Moon for hours.

Round craters, high mountains, and deep valleys are all on display whenever you look at the Moon through a small telescope. Observing the Moon through a telescope is best done during the crescent, quarter, or gibbous phases. You will want to focus your attention on the dividing line between the light and dark portions of the Moon. This line is called the terminator and that is where you will want to look to see the most dramatic plays of light and shadow on the uneven lunar landscape. The Moon's surface may look totally fake—like it is a bad movie set or made of clay. But that is the real Moon, up close and personal.

The Naked-Eye Planets

The ancients had a much different idea about the planets than we do today. They could not picture them as round rocky or gaseous objects that circle the Sun. They were not worlds or places that could even be visited. They were simply unique pinpoints of light that shone from an unfathomable distance. Or else they were gods who roamed and ruled the heavens above. Without a telescope, there was no way for an ancient astronomer to get a closer look at the planets, but there was one characteristic that made the planets totally fascinating: they wandered.

To the naked eye of these ancient stargazers, the planets seemed to be very strange-moving, sometimes suspiciously bright stars. In fact, the word *planet* comes from the ancient Greek phrase *aster planetes*, which means "wandering star." While all of the constellations remained fixed in their recognizable formations, there were seven exceptions to this rule. They wandered from place to place over days, months, and years.

We have already talked about two ancient "planets": the Sun and Moon. Since the Sun and Moon both appeared to wander across the background stars and constellations, they were considered to be planets by most ancient civilizations. The Sun moved slowly (about 1 degree per day), while the Moon took big jumps from night to night (about 13 degrees per night). But there are five other wandering planets that amazed, perplexed, and inspired the ancient stargazer—the five planets that you can see with the naked eye: Mercury, Venus, Mars, Jupiter, and Saturn.

Let's observe these five naked-eye planets through the eyes of the ancients. What made the planets unique? How did they move? What characteristics did each planet evoke? Then let's take some closer-up views of these wondrous objects and share something every ancient astronomer would be envious to see: a view of each planet through a small telescope.

MERCURY

What Is It?
Planet

Difficulty Level
Moderate

Description
Each planet has a personality. Each wandering star wanders through the constellations in a unique fashion. Mercury, associated with the swift, fleet-footed messenger god, lives up to the reputation of its namesake. It is the fastest planet, whipping around the Sun at an average speed of about 170,000 kilometers per hour.

Mercury is the most elusive planet and is probably the toughest of the five naked-eye planets to find in the night sky. It moves quickly from night to night, and it shifts from being visible in the morning sky to the evening sky in a matter of weeks. It took a long time for the ancients to figure out what was going on up there. In fact, some cultures considered Mercury to be two separate objects—a morning deity and an evening deity.

Best Times to See Mercury
The main problem with finding Mercury is that it orbits very near to the Sun. That means Mercury is most often up in the sky at the same time as the Sun. Sunlight washes out Mercury's feeble light, and this makes it invisible to the naked eye during most of the day. The only times you can actually find Mercury are when it appears farthest from the Sun (astronomers call this position of a planet its greatest elongation) while the Sun is still below the horizon. Sometimes you can catch Mercury just after sunset and other times just before sunrise. But without a telescope, you can never see Mercury in the middle of the day (the Sun is too bright) or late at night (because Mercury sets soon after the Sun and is below the horizon).

Here's where it gets complicated. Some elongations are better than others and some seasons make it easier to catch Mercury than others. Each year you get about three chances to see Mercury in the predawn sky and three chances to see it at twilight. In the Southern Hemisphere, the best months to see Mercury in the evening are September, October, and November, and it can be easier to spot the elusive planet before sunrise in March, April, and May. Your window to see Mercury at its greatest elongations (morning or evening) is small. You can only see Mercury far enough from the Sun for about a week around each elongation. Consult sky simulation apps or astronomy websites to find out when Mercury will be at its most favorable elongations for the year.

To see Mercury when it is at its greatest eastern elongation, you will want to look to the western sky 15 minutes after sunset. It will be low in the sky, so you will need a clear view to the western horizon free from buildings or trees. As the sky darkens search for a steady, semi-bright light glowing through the twilight. It will be about as bright as the brightest nighttime stars. But since it will be low in the sky, you might notice that it has a different color. Most observers describe Mercury as having a pinkish hue. The planet itself is actually dark gray, but the pink color shines for the same reason that a setting Sun turns from yellow to that memorable shade of red. The light bouncing off Mercury must travel through more layers of Earth's atmosphere to reach your eyes. This makes the light reflected off Mercury's surface appear a tinted pink.

To observe Mercury in the morning sky, you will have to wait for its greatest western elongation. When this occurs, look low above the eastern horizon 45 minutes before sunrise. Over the next 30 minutes you might see pinkish Mercury pop into view just before the Sun pokes above the horizon.

See Mercury Through a Telescope

Mercury's complicated relationship with the Sun makes this planet a challenge to observe clearly through a telescope. On those rare occasions when Mercury is visible and you can point a telescope toward it, your view may be disappointing. "Oh, that's it?" is a very common reaction to seeing the closest planet to the Sun through a telescope. First, Mercury is a very small planet—the smallest planet in the solar system—at only 4,879 kilometers in diameter (Earth's diameter is almost 13,000 kilometers). Plus, when you see Mercury it could be 80–160 million kilometers away. As a result, it will often look like a pinkish pinpoint of light even when you are viewing through a good telescope.

However, if you look closely you may notice that it is not a pinpoint or a perfect circle. Mercury can exhibit many shapes. At elongation Mercury should look like a tiny half-moon. And if you're really lucky and have a nice clear sky, it can sometimes appear as a little crescent. Why is this happening? Why does Mercury go through phases? Just like the Moon, Mercury gets all of its light from the Sun. You can observe different phases of Mercury depending on where it is in its orbit around the Sun. When it is at elongation, your perspective on the little planet only allows you to see half of it lit up. When it comes nearly between the Earth and Sun, Mercury looks like a little crescent.

Witness a Transit of Mercury

Since Mercury is the closest planet to orbit the Sun, does it ever come between the Sun and Earth? Can Mercury eclipse the Sun? Yes and no.

Yes, Mercury can pass directly between the Sun and Earth. But no, it is too small in our sky to completely eclipse the Sun. It will merely look like a tiny disc, a circular black blemish before the dazzling surface of the Sun. Astronomers call this a transit. Transits of Mercury are rarer than eclipses and occur only about thirteen times per century. The next transits of Mercury will be November 11, 2019, and November 13, 2032.

To observe a transit of Mercury, you will need to take the same precautions as you do when viewing the Sun. Protect your eyes by employing approved solar filters. And since Mercury seems so tiny in the sky, you will not be able to see this transit with the naked eye. You will need to magnify the image and outfit binoculars or telescopes with safe solar filters to see the alignment of the Sun and Mercury.

VENUS

What Is It?
Planet

Difficulty Level
Easy

Description
Venus was the goddess of beauty in ancient Rome and her brightly shining planetary namesake lives up to that hype. When Venus is in the evening sky, you notice her. When she is in the predawn sky, you can't miss her. When Venus shines, she looks like an extraterrestrial visitor who has graced the heavens. There is something UFO-like about her.

As the Sun dips below the western horizon and twilight descends into a deeper shade of blue, one "star" bursts into view before any other. That "star" is the planet Venus. It is so incredibly bright in the sky because a thick blanket of clouds perpetually covers its surface. These clouds reflect so much sunlight into space that from Earth only the Sun and Moon shine brighter than Venus.

Venus is the second closest planet to the Sun. Like the other inner planet, Mercury, Venus can best be seen just before sunrise or just after sunset, depending on where it is in its orbit. Some people call Venus the Morning Star or Evening Star, depending on when it is visible. Venus looks especially good when it is near a Waxing Crescent Moon in the fading twilight of evening, or when it cozies up to a Waning Crescent Moon just before dawn.

The Phases of Venus
Venus's brilliance comes at a price. The clouds that reflect so much sunlight completely obscure our view of the surface below. However, Venus presents a better sight through a telescope than Mercury. At 12,104 kilometers in diameter, Venus is a much larger planet than Mercury, and it can come much closer to Earth. So you can easily notice Venus's shape even through a small telescope.

When Venus is on the other side of the solar system, it looks like a small disc. It's almost a perfect circle. But as Venus rounds the Sun and starts coming closer to Earth, it changes phases: first into a gibbous Venus (which is almost full), then into a half Venus, and then into a crescent Venus. Since Venus only shines from reflected sunlight, you can see different phases of Venus, depending on where it is in its orbit around the Sun.

Also, Venus's apparent size changes dramatically. At its farthest, Venus is about 260 million kilometers from Earth. But when you see it as a slim crescent, it is only about 65 million kilometers away. So Venus can look four times larger just before coming between us and the Sun. In fact, during the crescent-Venus stages, the planet is so close to us that some people can even discern its shape with the naked eye.

See Venus in the Daytime

If you're looking for a challenge, try finding Venus during the daytime. During the several months when Venus may be visible in the evening sky right after sunset, strive to spot it before the Sun sets. During greatest elongation, when Venus appears to be farthest from the Sun, this is definitely possible. In fact, if the western sky is really clear, during the greatest elongations you can find Venus 30 minutes before sunset.

A tougher challenge is to find Venus with the naked eye during the middle of the day. The bright sunlight washes out the light of every star, and so they are impossible to find during the daytime. However, Venus is bright enough that if you know exactly where to look, you can find it as a pale white dot amid the light blue midday sky. It is extremely difficult to locate Venus this way but the Moon can help.

The easiest times to find Venus in the daytime are during Moon–Venus conjunctions. These happen when the crescent Moon appears near Venus in the sky. Sky simulation software or phone apps can show you when and where the Moon and Venus will be in conjunction and how they'll appear (they reveal whether Venus will be on the left or right of the Moon, for example). Then on the appropriate day, first find the

Moon in the sky. Once you do that, slowly scan the sky around the Moon until you spot a little pale dot. If you see that dot, it has to be Venus, because other than the Sun and Moon you cannot make out any other planet or star in the daytime sky. It takes some practice and patience, but once you get the hang of it you will be able to find Venus in the daytime.

Venus's Eight-Year Cycle

The ancient Mayan astronomers were obsessed with Venus. They charted its wanderings across the heavens with such precision that they could predict where Venus would be in the sky years in advance. From their countless observations, they discovered this pattern: Venus was visible in the morning sky just before sunrise for about 260 straight days. Then Venus went behind the Sun (from their perspective) and was not visible in the sky for about fifty-six days. At the end of that period, Venus popped out into the evening sky and remained visible each night after sunset for another 260 days. But then something strange happened: when Venus came between Earth and the Sun, it was only invisible for a scant eight days before reappearing again in the morning sky for another 260 days. They observed that the cycle repeated itself every 584 days.

The Maya also figured out that over a period of eight years, this 584-day cycle repeated itself exactly five times. If you trace Venus's path in the sky at the same time every day, you'll notice that it makes a weird loop. And if you chart the course of Venus over eight years, you'll record five distinct patterns: a squiggle down, a loop down, a zigzag, a loop up, and a squiggle up. That's five shapes that repeat almost exactly every eight years.

The Mayan astronomers demonstrated their knowledge of Venus in a book of tables called the Dresden Codex. Their Venus tables were so accurate that after 500 years, their predicted positions of Venus would only be off by one day. You can watch Venus just as carefully as the Maya. Note what time of day you see it, how high in the sky it is, and where it rises or sets. Do that for eight years and you'll have the pattern down pat!

MARS

What Is It?
Planet

Difficulty Level
Easy

Description
Mars is a small planet that is only 6,779 kilometers wide (about half the diameter of Earth), but it is still extremely easy to find in the night sky with the naked eye. It is nicknamed the Red Planet because it shines with an extremely off-white light. When you find it, you might classify its hue as orange or yellow instead of red, but when you compare its light to that of white or blue stars, you'll see it is redder than most. This bloody color led ancient Romans to associate this planet with Mars, their god of war.

In the sky, Mars generally appears brighter than the brightest stars. Even when very far from the Earth, Mars is on par with the first magnitude stars (the brightest stars) that are visible from even urban locations. But when Mars is especially close to Earth it is a dazzling orange beacon in the night. When Mars is nearby, only Venus and Jupiter can shine brighter.

Mars in Opposition
Mars is best viewed when it is closest to the Earth, when it appears in the evening sky at its biggest and brightest. The Red Planet's distance from us varies significantly: from almost 400 million kilometers at its farthest to about 56 million kilometers at its closest. Astronomers use the term *opposition* to describe Mars's regular close passes to Earth. Opposition occurs when the Earth is between Mars and the Sun, placing the Sun on the opposite side of the sky from Mars. This friendly alignment happens about every twenty-six months.

Not all oppositions are created equal. Because Mars has an eccentric orbit, some oppositions are much closer than others. At the 2003 opposition Mars was about as close as it could possibly get to Earth: 55.5 million kilometers. Other oppositions place Mars at 70, 80, or even 90 million kilometers away. Each opposition provides a better opportunity to check out Mars more closely, and the next exceptionally close oppositions will be in July 2018 and September 2035.

As the Sun sets in the west during opposition, you can spy the Red Planet rising in the east and blazing in the evening sky with a steady ruddy glow. Mars's dates of opposition and the constellation in which you can find it from 2018–2029 are:

- July 27, 2018, in Capricornus
- October 13, 2020, in Pisces
- December 8, 2022, in Taurus
- January 16, 2025, in Gemini
- February 19, 2027, in Leo
- March 25, 2029, in Virgo

If you miss seeing Mars on those dates, don't worry. Those are just technically the best dates to observe Mars. Mars will still be almost as bright for about a month before and a month after opposition.

After opposition, Mars tends to hang around in the evening sky for a long time. You will see it drifting slowly to the west, night after night, for six to nine months after opposition. It barely changes positions from night to night. This happens because Earth and Mars are traveling around the Sun in the same counterclockwise direction. Because Earth is on the inside track, it is moving faster than Mars, and so each night Mars drifts farther and farther away and appears dimmer and dimmer in the night sky.

The Peculiar Motion of Mars

Like all planets, Mars wanders across the night sky, appearing to travel from constellation to constellation. However, it does not keep up a steady pace. It goes along, slows down, stops, reverses itself, slows down again, and then goes forward once more. This pattern repeats every twenty-six months. If you took a picture of Mars every night at the same time, you would see that it makes a loop-de-loop pattern. So what is going on here? This is Mars's retrograde motion, and although this distinctive motion was well known to ancient astronomers thousands of years ago, it took until the 1500s for a Polish astronomer to correctly explain it.

The ancient models of the universe placed an unmoving Earth at the center surrounded by the circular pathways of the Sun, Moon, and planets (Mercury, Venus, Mars, Jupiter, and Saturn). But if the planets just circled Earth, why did Mars make a loop-de-loop exactly when the Sun was on the other side of Earth? Ancient astronomers scrambled for answers, and they came up with a mechanism called an epicycle. Basically, they proposed that Mars completed a circle upon a circle. Eventually the astronomers needed epicycles upon epicycles to make their models match what appeared in the sky above. It became really, really complicated, but it worked for over a millennium.

When the Polish mathematician and astronomer Nicolaus Copernicus checked on the planets in the early 1500s, he discovered that Mars and Saturn weren't where they were supposed to be. He thought that the epicycles were too complicated and obviously flawed, so Copernicus began searching for an alternative explanation. After intently studying the heavens, he explored what would happen if the Sun was at the center of the universe instead of Earth. Copernicus then hypothesized that Earth and Mars both travel around the Sun, but Earth moves faster. As it catches up to Mars and passes it, the Red Planet appears to move backward in the sky. The loop-de-loop was just part of a grand illusion. Earth actually moved! Copernicus's Sun-centered universe was perhaps one of the greatest discoveries of all time, and it was the motion of Mars that provided the biggest clue.

Mars in a Telescope

You can always find Mars near one of the twelve zodiac constellations (e.g., Leo and Pisces). Many star charts and astronomy apps will tell you in which constellation to look for Mars and what time of night it is easiest to see. Before it reaches opposition, you can find Mars in the predawn sky in the south. During opposition it rises in the east just after sunset. And after opposition Mars hangs out in the southern sky after dark. No matter the season, if Mars is in the sky it will be brighter than almost every star in the surrounding constellation in which it resides.

When you find Mars in a telescope, do not expect to see anything nearly as exciting as Martians. Although Mars may be the most fascinating planet, through a backyard telescope it can appear underwhelming. Mars is a small planet and it looks like a small orange disc through most telescopes. However, when Mars is closest to Earth you may be able to detect features on the Martian surface.

If the planet is properly tilted, you may be able to discern a white spot marking a polar ice cap. At other latitudes on the surface you may see darker markings such as lines, blobs, and triangles. These were known as the Martian canals in the nineteenth and early twentieth centuries. We now know that these so-called canals are merely darker-colored rocks that show up against the rusty soil of Mars.

The most dramatic Martian feature you can see through a telescope is called Syrtis Major. It looks like a dark brown triangle or chevron on the orange surface of Mars, but it is in fact the remnants of a shield volcano that has been dormant for eons.

Whether you observe Mars with the naked eye, binoculars, or a telescope, it may inspire you to dream of Martians visiting Earth someday or of Earthlings crossing the millions of kilometers of space to colonize a new planet. Let your imagination go wild under the glow of the Red Planet.

JUPITER

What Is It?
Planet

Difficulty Level
Easy

Description
Ancients Greeks called it Zeus, and the Romans adopted this bright night-light as the manifestation of their chief god, Jupiter. How did the ancients know Jupiter was the king of the planets? It's not the brightest planet—that's Venus. It's not the fastest- or slowest-moving planet—that's Mercury and Saturn, respectively. There must be something kingly about Jupiter that was visible even to the naked eye.

Jupiter is by far the largest planet in our solar system. It has an equatorial diameter of about 140,000 kilometers, meaning that 1,325 Earths could fit inside it. In fact, Jupiter is more massive than all of the planets, moons, and asteroids in the solar system combined! So its name really fits what we know about this giant planet.

Identifying Jupiter
Jupiter is an unmistakable light in the night sky. It appears to be a non-twinkling cream-colored star and is very often the brightest starlike object in the entire night sky. Jupiter's brilliance is so stunning that it is often more than twice as bright as the brightest star in the sky, Sirius. When Jupiter is in the sky, you notice it!

Jupiter is a steady performer. The light it shines on Earth does not fluctuate very noticeably in brightness, like Mercury or Mars, over the time it is visible in the night sky. Even though Jupiter is far from Earth (roughly 600 million kilometers when it is "close"), it is so large and its cloud tops reflect so much sunlight that it shines brightly whenever it is up in the sky. Even when it is on the other side of the solar system and slightly farther from Earth, Jupiter dazzles.

Like all planets, you can always find Jupiter hanging out among the stars of the zodiac constellations. It wanders among the constellations much more slowly than any planet we have studied so far. In fact, it takes Jupiter 11.86 Earth years to orbit the Sun. That means each year Jupiter shifts its position in the sky by one-twelfth of a circle. This circle is called the zodiac, and Jupiter visits each zodiac constellation in order, one by one, year after year. For instance, if you see Jupiter in front of the stars of Libra, the next year it will be in front of Scorpius. The year after that it will grace the presence of Sagittarius, and so on. About every twelve years Jupiter will return to nearly the same location in the sky, completing a grand tour of the zodiac. Check astronomy websites and star charts to know in which zodiac constellation to look for Jupiter this year.

Jupiter at Its Brightest

Like Mars, Jupiter is closest to Earth near its opposition. Jupiter takes almost twelve years to revolve around the Sun, and so Earth catches up to it about every thirteen months. Thus, oppositions, and the best season for Jupiter-viewing, occur about thirteen months apart.

The oppositions of Jupiter between 2019 and 2030 are listed here. You don't have to look for Jupiter only on these dates, however. You can find the giant planet shining brightly for several months before and after opposition. Astronomers just like to be technical and give you the absolute best nights to observe Jupiter and where to look, which are:

- June 10, 2019, between Ophiuchus and Scorpius
- July 14, 2020, in Sagittarius
- August 20, 2021, between Capricornus and Aquarius
- September 26, 2022, in Pisces
- November 3, 2023, in Aries
- December 7, 2024, in Taurus
- January 10, 2026, in Gemini
- February 11, 2027, between Cancer and Leo
- March 12, 2028, in Leo
- April 12, 2029, in Virgo
- May 13, 2030, in Libra again

Jupiter Through Binoculars

If you view Jupiter through a pair of binoculars, you may be in for a special treat. Although you may not be able to see any features on the planet itself, you should be able to pick out a few little dots next to the planet. These are Jupiter's largest moons, Io, Europa, Ganymede, and Callisto. There are some rare humans who possess seemingly superhuman eyesight and can see these four moons with their naked eyes. Test your eyesight and see if you can do it too. If not, get some binoculars—or, better yet, a telescope.

Jupiter Through a Telescope

With even a small telescope, Jupiter looks amazing. You can re-create the observations of Italian astronomer Galileo Galilei, who turned his homemade telescope toward Jupiter in 1610. Galileo saw Jupiter's four largest moons (now called the Galilean Moons) and computed their orbits. Each night Io, Europa, Ganymede, and Callisto form a different pattern. One night there are two on one side and two on the other. Another night you might see three on one side and only one on the other. And sometimes you'll only see three moons because the other is hiding in front of or behind the planet.

Observe Jupiter and its moons every night for a week or two and see if you can figure out the pattern to their motions. How can you even tell which is which? Somehow, Galileo could tell them apart more than 400 years ago.

Through a midsize or a large telescope Jupiter really comes to life. You can see the disc of the planet, two major dark bands, several smaller bands, and, sometimes, the Great Red Spot. This persistent mega-hurricane on Jupiter is always located in one of the thicker dark bands that ring the planet. You may first note the Great Red Spot's presence as a break in one of the bands, like a chunk is missing from it. But as you look more closely, you may be able to see this circular spot that is in reality much larger than Earth.

When one of the Galilean Moons comes between the Sun and Jupiter, you can tell. The moon will cast a shadow onto the surface of the planet. It looks like a little black spot on Jupiter. Occasionally you can catch two or even three circular shadows on Jupiter at one time. As you can see, Jupiter truly is one of our most remarkable planets.

SATURN

What Is It?

Planet

Difficulty Level

Easy

Description

As "wandering stars" go, Saturn is the slowest-moving planet. Whereas Mercury whips around the Sun at over 170,000 kilometers per hour, Saturn practically pokes along at about 35,000 kilometers per hour. That means that although Saturn wanders across the background stars like its fellow planets, it does so at an incredibly slow pace.

It takes weeks, months, or even years to note any change in Saturn's position relative to the background constellations. While Earth circles the Sun every year, it takes Saturn about 29.5 years to complete one orbit. If you observe Saturn one starry night and come back to it one year later, the stars will be in the same place, but Saturn will have moved only about 12 degrees to the east. That means if you see Saturn among the stars of Sagittarius it can take two to three years for Saturn to move on to the next constellation of Capricornus.

The ancient Greeks and Romans noticed this sluggish motion and incorporated it into their mythology. Saturn (also known as Cronus in Greece) was the father of Jupiter (Zeus). After Jupiter overthrew his father and became the supreme god, Saturn became an old man and receded into the background. Saturn is often depicted as an old man with a long beard and later was equated with the figure of Father Time.

Identifying Saturn

The farthest planet you can see with the naked eye is Saturn. Although you cannot detect the ridiculously cool rings of this planet without a telescope, you can still easily locate it every year in the night sky.

Saturn appears to be a non-twinkling yellow star that shines with a light equal to or sometimes greater than the brightest first magnitude stars like Vega and Arcturus. Saturn definitely does not stand out as much as Venus, Mars, or Jupiter, but once you find it you will be able to go back to it night after night since it changes positions so slowly. Saturn never strays far from the zodiac constellations, so you can check astronomy websites or sky simulation apps to discover what area of the sky the ringed planet will be in for more than a year in advance.

Saturn at Its Brightest

Like all planets farther from the Sun than Earth, Saturn appears biggest and brightest near opposition. At opposition Saturn is still over one billion kilometers away, but it is so large and its surface and rings reflect so much sunlight that it shines like a first magnitude star.

Since Saturn is so slow-moving, Earth can round the Sun and catch up to it almost every year. The following list contains the opposition dates for Saturn between 2019 and 2030 and in which constellation you can find it. Notice that there is only an eleven- or twelve-day difference in dates from year to year. Just like Jupiter, you can see Saturn shining nearly as brightly up to one month before and one month after opposition. The dates are:

- July 9, 2019, in Sagittarius
- July 20, 2020, in Sagittarius
- August 2, 2021, in Capricornus
- August 14, 2022, in Capricornus
- August 27, 2023, in Aquarius
- September 8, 2024, in Aquarius
- September 21, 2025, in Pisces
- October 4, 2026, in Pisces
- October 18, 2027, in Pisces
- October 30, 2028, in Aries
- November 13, 2029, in Taurus
- November 27, 2030, in Taurus

Saturn Shines in a Telescope

What are the best things to see in a telescope? Saturn, Saturn, and Saturn.

It is a rite of passage to locate Saturn and observe it in a telescope. When you swing a telescope toward it, center it, and place your eye to the eyepiece, something magical will happen. There Saturn will sit among the blackness of space, a tiny cartoon world encircled by rings. It will look totally fake—as if someone placed a sticker of Saturn at the end of your telescope. But it is the real thing! You are experiencing sunlight bouncing off an improbable planet almost one billion kilometers away, coming through the telescope and into your eye.

Saturn has a transformative power that excites our imagination like no other astronomical object. Saturn is our childhood symbol for space, and seeing the real thing literally lights up our faces.

Most telescopes can allow you to detect several of Saturn's moons. Titan is by far the brightest and will look like a steady-shining star off to one side of the planet. The moons Rhea, Tethys, and Dione can be spotted with a moderate telescope. If you're lucky enough to view Saturn under ideal conditions through a large telescope, you may be able to make out the fainter moons Iapetus, Enceladus, and Mimas as well.

Ring Tilts and Saturn Seasons

Every year Saturn looks a little different in a telescope. During its 29.5-year orbit of the Sun it's tilted by 27 degrees with respect to its orbit around the Sun. This slant changes your viewing perspective on the rings. When the rings are tipped up or down to you, Saturn reveals the classic views you're used to. When they are tilted to the max, you can behold the breadth of the rings, observe their structure and gaps, and witness how they cast shadows onto the planet.

The rings are surprisingly svelte, and when they close down and become edge on to you, they become incredibly difficult to see. In some places Saturn's rings are barely 100 meters deep. That means when the plane of the rings is pointing directly at Earth, they are invisible through even the most powerful telescopes. When Saturn seems to be missing its rings, astronomers sometimes call this "naked" Saturn. The next naked Saturn will occur in 2025, but until then make it your goal to see Saturn in a telescope. You will not be disappointed.

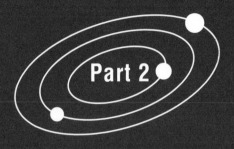

STARS AND CONSTELLATIONS

Ancient stargazers marveled at all of the lights above—the different luminosities, shades, shapes, and bands of stars. The stars were gods, unreachable but ever-present with hints of personality. Stargazers watched how the stars rose, crossed the sky, and set. Some stars even fluctuated in brightness and brought intrigue to the masses. Ancient stargazers from Egypt to South Africa, Australia to New Zealand, and the Americas to Europe, connected the dots in the sky and formed pictures called constellations.

You are now going to take up the pastime of your ancestors. You will hear the stories and look for the same stars they observed thousands of years ago. The positions and brightness of the vast majority of stars have not changed significantly in the last four millennia, and as a result the outlines of the constellations have not drastically altered. So you will essentially be seeing the same sky with the same configurations of heavenly lights that have glittered and glistened on every great civilization throughout human history.

In this part I've broken the sky down into five sections. First, we will explore the southern sky and stars that are visible almost all year round from the mid-Southern Hemisphere. Then we will take a closer look at the stars and constellations visible by season: the summer sky, autumn sky, winter sky, and spring sky. Each section will start with a wide-angle view of the firmament in order to get a better overview. Then we will zoom into individual constellations and visit the most interesting stars that reside within them. Since timing is important here, please note that the charts display the view of the sky during evening hours—prime-time viewing for most people. Let's head to the sky!

The Southern Sky

For sky watchers living in the Southern Hemisphere, the stars in the southern sky hold a special place of honor. Over the course of every night these stars rotate around a central point. After an entire day each southern star makes a complete circle and returns to its starting point. These stars are called circumpolar stars, meaning they circle the South Celestial Pole and never set. They are visible every night of the year.

From the midlatitudes (like Australia, South Africa, and Patagonia) the circumpolar constellations include Crux, the Southern Cross; Triangulum Australe; Hydrus, the Water Snake; and Pavo, the Peacock. At the very center of the circumpolar stars, at the seeming pivot point of the southern sky, lies a faint constellation called Octans.

Picture the stars and constellations in the southern sky as if they were etched into a circular plate. These shapes seem fixed onto this plate, and once a day the plate spins clockwise around Octans. Octans is at the center of the plate and makes a tiny circle, barely moving. Since Octans has no bright stars and may be completely invisible from urban locations, southern stargazers are greeted with a seemingly empty space. But there are other tricks to finding direction from other circumpolar stars and Crux, the Southern Cross will help guide you around the southern sky.

When you watch the southern sky for an hour or two, you can very easily detect this clockwise motion of the stars around a central point. If you set up your camera on a tripod and take a very long exposure, you can capture star trails—arcs traced by the light of the stars moving around the space where Octans resides. This grand circle is caused by the rotation of Earth, the same motion that makes the Sun rise and set. This celestial circumnavigation is marvelous to experience under a dark sky. Let's take a look at these southern sky stars and constellations.

CRUX, THE SOUTHERN CROSS

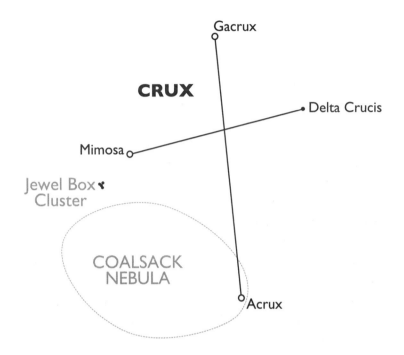

What Is It?
Constellation

Difficulty Level
Easy

Description
Whether you live in the heart of a city and see a small faint cross shape of stars above the skyline or you live in the country and see this shape embedded in the river of stars called the Milky Way, the constellation Crux or the Southern Cross is the landmark of the southern sky.

Crux is the smallest constellation in the entire sky. The longest part of the Cross, from its stars Acrux to Gacrux, is 6 degrees, and the shorter axis, from the stars Mimosa to Delta Crucis, is only 4 degrees wide. Despite its small size, Crux contains stunning stars and deep space delights like the Jewel Box cluster and the Coalsack Nebula. To name

Crux's stars, start at the bottom of the cross to find bright Acrux and moving clockwise around the star pattern you will come to Mimosa, Gacrux, and finally Delta Crucis.

To some Australian Aboriginal peoples, the Southern Cross represented a possum sitting in a tree. Maori people called it the anchor. In Brazil, Indonesia, and Malaysia the stars formed the body of a ray. In Samoa it was a triggerfish followed by two men (the stars Alpha Centauri and Hadar in the constellation Centaurus). And on the islands of Tonga it was known as the duck with wings spread and flying south.

In addition, not only is the image of the Southern Cross affixed to the flags of Australia, New Zealand, Papua New Guinea, Samoa, and Brazil, but it is also mentioned glowingly in the national anthems of Australia and Brazil.

Also, although there is no "South Star," navigators use the Southern Cross to help tell directions. Connect the dots of the longer axis of the cross (from Gacrux to Acrux) and continue that line of sight for about another 25 degrees. That will take you to the point in the heavens around which the southern sky rotates daily. That direction in the sky is south. Another trick is to make a fist and extend your arm. Place the knuckle of your little finger over Gacrux and the knuckle of your pointer finger on Acrux. When you extend your thumb, it will inevitably point to the southern horizon. With this info you can tell your directions.

How to Find It

Crux is the easiest constellation to find in the southern sky and is visible nine months out of the year. Face south and you will almost certainly see the four stars making a cross or kite-shaped pattern. It stands highest in the southern sky on April, May, and June evenings and can be between 50 and 70 degrees above the horizon (depending on your latitude). In July and August the cross is halfway up in the south-southwest and by September can be very low in the south just after dark. October through December is the toughest time to spot the cross since it can be below the horizon from most southern latitudes. But January–March it rises up in the south-southeastern sky every night.

ACRUX, MIMOSA, GACRUX, AND DELTA CRUCIS

What Is It?
Stars

Difficulty Level
Easy

Description
These are the four bright stars that form the cross-shaped pattern of the constellation Crux, the most prominent little star pattern in the entire southern sky. They are often labeled with Greek letters to indicate their brightness (Alpha, Beta, Gamma, and Delta). But their common names (in order of brightness) are Acrux, Mimosa (sometimes called Becrux), Gacrux, and Delta Crucis.

Although the stars look close together they lie at various distances from Earth. Acrux is a blue star 321 light-years away. Mimosa is another blue-hued star but is slightly closer at about 280 light-years. Gacrux is a red giant and closest of the four stars in the Southern Cross at only 88 light-years. And Delta Crucis, the dimmest and farthest of the four, is blue in color and lies about 345 light-years from Earth.

How to Find It
Identifying these four stars is about as easy as it gets in astronomy. Simply locate the constellation Crux, the Southern Cross and you've found them. Three of the four stars are of first magnitude and are so bright that they can be seen even from light-polluted skies. The fourth star (Delta Crucis) is of second magnitude and can be easily spotted from suburban and rural locations.

The stars in Crux are circumpolar and can be found at some point in the night during all months of the year. They sit high in the southern sky during April, May, and June evenings and by July, August, and September are about halfway up in the southwest. October, November, and December are the most difficult evenings to look for these stars since they are low along or below the southern horizon. But by January, February, and March they start rising higher in the southeastern sky.

THE JEWEL BOX

What Is It?
Star cluster

Difficulty Level
Moderate

Description
The Jewel Box in the constellation Crux, aka the Southern Cross, is an open cluster of about 100 stars. It got its flashy name from the English astronomer John Herschel, who, after seeing it in his telescope, described the glittering stars as "a superb piece of fancy jewelry."

At about 14 million years old, the stars within the Jewel Box are some of the most recently formed stars in the galaxy. The cluster may be difficult to see in light-polluted areas since it shines, as a whole, with the equivalent brightness of a fourth magnitude star (more than a dozen times fainter than Acrux). If the Jewel Box were closer to Earth it would be even more stunning. Unfortunately it lies about 6,400 light-years from Earth, making it one of the farthest things in our galaxy you can see with the naked eye.

How to Find It
The Jewel Box cluster is observable almost every month of the year in the southern sky. Only on November and December evenings can it be so low in the sky as to make viewing difficult. If you can see the Southern Cross, you can find this cluster.

The Jewel Box appears very close to the second brightest star in Crux, Mimosa. In fact, it's only about 1 degree (smaller than the width of your little finger at arm's length) from this bright and notable star. If you're still having trouble locating this cluster's position, connect a line between the top star in the Southern Cross, Gacrux, and Mimosa (the left side of the cross) and continue that line 1 degree farther. If you are viewing the Jewel Box cluster with the naked eye from a dark or semi-dark sky you should be able to see it as a slightly fuzzy star. If you're viewing it through a pair of binoculars or a small telescope, this star cluster will be much clearer and you can behold the dazzling jewels. See if you think it sparkles like diamonds among the black backdrop of the sky.

COALSACK NEBULA

What Is It?
Dark nebula

Difficulty Level
Moderate

Description
Ninety-nine of the 100 things to see in this book are visible because they are lit by stars. The Coalsack Nebula 600, which is light-years away, is the exception. This dark nebula of cosmic dust is so huge that it blocks the light of the stars behind it and looks like an irregularly shaped black hole in front of the more distant Milky Way.

Australian Aboriginal mythology describes the Coalsack Nebula as the head of the great Emu constellation. Instead of using stars to depict their star creatures, Aborigines used the dark and dusty lanes of the Milky Way galaxy in addition to the Coalsack Nebula to form an Emu in the heavens.

An Incan legend from Peru describes the Coalsack as the place where the god Ataguchu kicked a hole in the Milky Way. The hole or divot in the sky became the Coalsack Nebula and the chunk that flew off went on to form the Small Magellanic Cloud (a galaxy you'll learn about later in this section).

How to Find It
To see the Coalsack at its best, you will need a sky dark enough to see the Milky Way—which will look like a band of milky clouds stretching through the constellation Crux. Find the two brightest stars in Crux (Acrux and Mimosa) and connect a line between them. As you travel from Acrux to Mimosa you may detect a vast empty area surrounded by the scattered starlight of the Milky Way; just look for a patch of sky 7 degrees long and 5 degrees wide that looks completely black and free from stars.

Since the Coalsack lies within Crux, like the constellation, it is also circumpolar and can be seen nearly all year round in the southern sky.

TRIANGULUM AUSTRALE,
THE SOUTHERN TRIANGLE

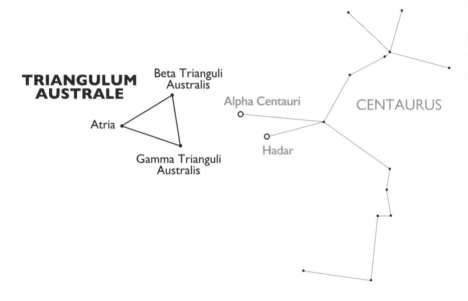

What Is It?
Constellation

Difficulty Level
Moderate

Description

The constellation Triangulum Australe may seem like a boring, uninspired star pattern with no extensive mythology. While it's true that there seem to be no strong, ancient legends about the stars that make up this constellation, after you identify it in the sky you have to admit that it really looks like a triangle. In fact, your imagination might be hard pressed to picture anything other than a triangle.

Triangulum Australe lies so far south in the sky that it was not visible from Europe or much of North America or Asia. However, after his voyage to the southern seas in the sixteenth century, Italian explorer Amerigo Vespucci described a triangle of stars in roughly this area of the sky. But the constellation we recognize today first appeared on a globe of the sky made by the German mapmaker Johann Bayer in 1603.

The main stars in this triangle pattern are called Atria, Beta Trianguli Australis, and Gamma Trianguli Australis. Atria (an abbreviation of "Alpha" and "Trianguli") is the brightest of the three and should appear slightly orange in color. It is a giant star with a diameter 130 times and a luminosity 5,500 times that of the Sun. If Atria were our star, its surface would engulf the orbits of Mercury and Venus.

How to Find It

The Southern Triangle is nestled in between four constellations so dim that they are not detailed in this book (Ara, Apus, Circinus, and Norma). But this star pattern is still easy to find because of its equilateral triangle shape and it proximity to two dazzling stars.

The easiest way to identify Triangulum Australe is to first find the brightest stars in the constellation Centaurus: Alpha Centauri and Hadar. From southern latitudes, Alpha Centauri and Hadar are the brightest pair of stars that appear close together nearly every night in the southern sky. Connect a line between Hadar (the dimmer one) to Alpha Centauri (the brighter one) and continue that line. Then curve the line slightly to the left and, after 9 degrees of sky, you will come to one corner star of the Southern Triangle (Beta Trianguli Australis). If you can see Beta you should be able to find Atria (7 degrees away) and Gamma (6 degrees away) to form a neat triangle.

Triangulum Australe is a circumpolar constellation and can be found every night of the year in the southern sky. In March, April, and May it is about halfway up in the south-southeast after dark. June through August you can find it due south and about two-thirds of the way up in the sky. September through November the Southern Triangle is halfway up in the south-southwest. And in December, January, and February it is low in the south after dark.

Do not mistake another triangle of stars called Hydrus for Triangulum Australe. Hydrus is formed from a larger triangle of fainter stars that we will cover in the following entry.

HYDRUS, THE WATER SNAKE

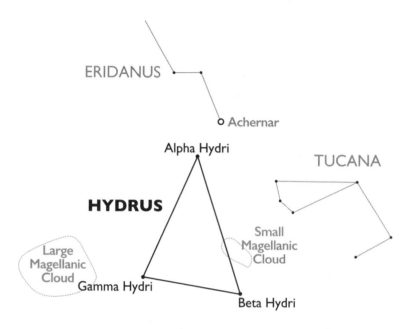

What Is It?
Constellation

Difficulty Level
Moderate

Description
In the night sky astronomers have designated a Northern and Southern Cross, a Northern and Southern Triangle, and a Northern and Southern Crown. Consider the constellation of Hydrus to be the Southern Snake, cousin to Hydra, the long and sinuous star pattern visible in the northern sky.

This snake is formed by three semi-bright stars: Alpha Hydri, Beta Hydri, and Gamma Hydri. The constellation was an invention of Dutch mapmaker Petrus Plancius who, in 1598, added its depiction to his star charts to fill in the space next to Tucana, the Toucan and below the river constellation Eridanus. Hydrus was described as a water snake and so its place at the end of Eridanus seemed a natural fit. It was also imagined to be a male water snake distinguishing it from the allegedly female Hydra, whom, according to Greek mythology, Hercules fought and killed. How did Hydrus get into the sky and what tall tales surround this snake? You will have to make them up yourself because by themselves, these stars have no long-lasting mythology.

The greatest claim to fame for this inconspicuous constellation is that it straddles two magnificent deep space objects: the Large and Small Magellanic Clouds. These dwarf galaxies consisting of millions to billions of stars appear like vast cloudy expanses in the southern sky. The Small Magellanic Cloud lies between Alpha and Beta Hydri while the Large Magellanic Cloud sits about 10 degrees to the east of the Gamma Hydri.

How to Find It

Hydrus can be confused with the smaller and brighter constellation of Triangulum Australe since they both have similar shapes. However, Triangulum Australe is about 40 degrees away. Some people actually look for both triangle-shaped star patterns (since they are often both visible at the same time), and note Hydrus as the fainter and slightly larger three-sided figure.

Hydrus is about halfway between the second brightest star in the entire sky, Canopus in the constellation Carina, and another bright star named Alpha Tucanae in Tucana. You can confirm the location of Hydrus by noting another bright star, Achernar, that is extremely close to the apex of one point of the triangle of Hydrus. If you connect the dots from the dimmest star in the triangle of Hydrus (Gamma Hydri) to Alpha Hydri and keep going, after another 5 degrees you will run into Achernar, which marks the southern terminus of the river constellation Eridanus (a constellation best seen in the summer months).

Hydrus is located so far south that it is visible all year at some point every night. During the evening hours look for a triangle shape of three dimmer stars halfway up in the southern sky from October through January. In February and March it hangs lower in the south-southwest. From April through July it is very low in the sky and is just above the southern horizon. In August and September it is about a third of the way up in the south-southeastern sky after dark.

THE MAGELLANIC CLOUDS

What Is It?
Galaxies

Difficulty Level
Easy

Description
Two stellar highlights of the Southern Hemisphere sky are the Large and Small Magellanic Clouds. They are irregular, dwarf galaxies or conglomerations of hundreds of millions to several billions of stars. Astronomers are not sure if the Magellanic Clouds are satellite galaxies of the larger Milky Way in which we live and are orbiting the Milky Way like moons circling a planet, or if they are merely passing through the neighborhood on their path through the larger universe.

The Magellanic Clouds get their name from the Spanish explorer Ferdinand Magellan who observed them during his voyage around the world from 1519–1522. You can see both Magellanic Clouds with the naked eye from locations close to or below the Equator.

As you'd expect, the Large Magellanic Cloud is the larger of the two galaxies and contains roughly 30 billion stars. It is a huge feature in the night sky even from its distance of 163,000 light-years from Earth. It still covers a swatch of sky almost 10 degrees in diameter (that's twenty times wider than the Moon appears in our sky).

While smaller than its counterpart, the Small Magellanic Cloud is still quite a sight to behold. Comprising hundreds of millions of stars that lie about 200,000 light-years away, it covers an area of the sky over 4 degrees wide (more than eight Moon-widths).

How to Find It
If you live in the heart of a city, the Magellanic Clouds may be impossible to see. They are about as bright as the Milky Way, so light pollution will definitely obscure them. But if you are viewing from a dark or semi-dark sky, don't worry, you can't miss them. Face south and look for two clouds of stars that look like pieces broken off of a long stretch of Milky Way.

They are both circumpolar and can be seen all year at various altitudes above the southern horizon.

The Small Magellanic Cloud is technically in the constellation Tucana but lies nearest to one side of the constellation Hydrus, the Water Snake, halfway between the stars Alpha and Beta Hydri. The Large Magellanic Cloud is embedded among four faint constellations: Dorado, Mensa, Reticulum, and Volans. Connect a line between Beta and Gamma Hydri and keep going another 8 degrees. There you will find the Large Magellanic Cloud.

PAVO, THE PEACOCK

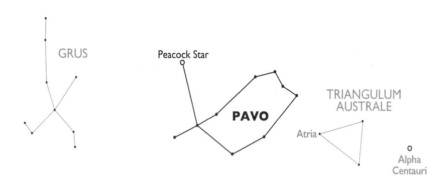

What Is It?
Constellation

Difficulty Level
Moderate

Description
Like many of the constellations that are visible mainly from the Southern Hemisphere, Pavo, the Peacock does not have a direct connection to ancient Greek astronomers living over 2,000 years ago. This star pattern was first designated as Pavo by Petrus Plancius, a Dutch astronomer, around 1598. Was Plancius inspired to create this constellation by a species of newly discovered peacocks during voyages of exploration? Was he inspired by mythology? No one is sure, but Plancius would certainly have been familiar with the following Greek myth involving a peacock that may have stoked his imagination.

Once upon a time, the chief god Zeus fell in love with a mortal woman named Io. To hide this maiden from his wife Hera, Zeus changed her into a heifer. Hera, knowing full well of this attempt at trickery, took the heifer into her special flock and guarded this nice gift from her cheating husband with Argus, the hundred-eyed monster. Argus would always keep at least one eye on Io and keep Zeus away.

Zeus teamed up with Hermes, the fleet-footed and fast-talking messenger god to help him get Io back. Hermes went to Argus and told a long and winding, dull and boring story to the hundred-eyed monster.

As time went on, Argus became sleepy and, one by one, Argus's eyes began to close. After the final eye shut and Argus was fast asleep, Hermes slew the beast and freed Io. Hera memorialized her monster by affixing his hundred eyes to the tail feathers of every peacock in creation.

It may be difficult to imagine a bird perched in the sky within this oval-shaped star pattern. The outline looks more like the head of an alien creature with two antennae sticking out. Furthermore, most of the stars in this constellation are of the third and fourth magnitude (meaning you need a semi-dark sky to even see them) and so they don't stand out. There is one exception: the brightest star in Pavo is of the first magnitude and is appropriately named the Peacock Star or simply, "Peacock." It's a blue-white star six times more massive than our Sun and lies about 177 light-years from Earth.

How to Find It

Pavo is moderately challenging to identify in the sky since it is difficult to make out its shape from urban locations. The best way to find it is to pinpoint its brightest star, Peacock, which lies about halfway between the nearby constellations of Grus, the Crane and Triangulum Australe, the Southern Triangle. All three constellations circle around the southern sky together.

You can also use the bright star Alpha Centauri and Triangulum Australe to guide you. From southern latitudes, the stars Alpha Centauri and Hadar are the brightest stars that appear close together nearly every night in the southern sky. Alpha Centauri is the brighter of the pair. Once you've located this star, make a line from Alpha Centauri and go through the center of Triangulum Australe for 15 degrees until you reach the star Atria, the brightest of the three stars in this triangle star pattern. Continue that line of sight (with a slight bend to the right) and after another 26 degrees, you will reach the Peacock.

Pavo is a circumpolar constellation and can be found in the southern sky nearly all year round. As winter begins, you can find the Peacock about halfway up in the southeastern sky every evening. During the autumn it rides high in the south and can be about 60 degrees above the southern horizon. By summer it circles over to the southwest and is much lower in the sky, about 30 degrees above the horizon. And during the autumn evenings you may have trouble seeing it since it sits just above the southern horizon.

TUCANA, THE TOUCAN

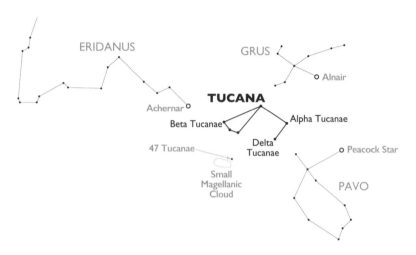

What Is It?
Constellation

Difficulty Level
Moderate

Description
As European explorers traveled south, they experienced different cultures and climates, unique wildlife, and brand-new stars (to them at least). Sixteenth-century Dutch mapmaker Petrus Plancius asked a few explorers to accurately map the southern stars they saw and bring the data back to him. From these, Plancius made maps of the newly discovered stars and invented twelve wholly new constellations.

You've already met Pavo, the Peacock, who borders Tucana to the west, but Plancius was fascinated with stories of the colorful South American bird and created the constellation Tucana to include on his star charts. The name and image of a tropical bird in the sky caught on and was included in celestial globes in the early seventeenth century. The Toucan is now firmly affixed in the sky and is officially recognized around the world.

The brightest star in Tucana is appropriately named Alpha Tucanae. It is a very cool orange-red-colored star the lies about 200 light-years from Earth. Alpha Tucanae is nearly a third magnitude star and is tough to find from urban locations. But the other stars in this constellation, such as Beta, Gamma, and Delta Tucanae, are all more than two times fainter, so the Toucan is best seen under dark skies.

How to Find It

The Toucan is moderately difficult to locate in the sky because it does not have many bright stars (and doesn't really look like a toucan). But you can find its position in the sky with the help of some brighter landmarks—and you can confirm its position with some breathtaking deep space objects that lie nearby.

Tucana lies between three bright stars: Achernar in the constellation Eridanus, the Peacock Star in Pavo, and Alnair in Grus, the Crane (whom you'll meet in the spring sky). Connect a line between Achernar and the Peacock Star, about 40 degrees in length, and you'll find the body of the Toucan halfway between the two stars.

Look for the nearby Small Magellanic Cloud (which lies mainly within the boundary of Tucana) and a tiny globe of stars called 47 Tucanae that looks like a piece of dust that fell out of this cloud. You may notice these out-of-this-world sights more than the Toucan and they can confirm the location of this more obscure constellation better than any other method.

You can find Tucana almost all year round. From September through December it perches 50–70 degrees high up in the southern sky. January, February, and March brings it lower in the southwest. April, May, and June evenings present the toughest months to see it in the evening since it sits low above the southern horizon. But come July and August, Tucana rises higher and is halfway up in the southeastern sky.

47 TUCANAE, LACAILLE'S TOUCAN

What Is It?

Globular cluster

Difficulty Level

Moderate

Description

Within the constellation Tucana you can find one of the most amazing clusters of stars in the night sky. It is called 47 Tucanae and it is the second brightest globular cluster in the Milky Way galaxy (behind Omega Centauri). Although it can be seen with the naked eye, the French astronomer Nicolas Louis de Lacaille is said to have "discovered" it during an expedition to the Cape of Good Hope, South Africa, in 1751. Lacaille noted 47 Tucanae when, not being familiar enough with the view of the heavens from the Southern Hemisphere, he mistook it for the nucleus of a comet. It is in fact an expansive cluster of more than 1 million stars—with a mass of about 700,000 Suns—that resides about 16,700 light-years from Earth.

At first glance this star cluster looks like a fuzzy star or a small cloud. But upon closer examination you should be able to make out 47 Tucanae's stellar nature. If you point a pair of binoculars at the cluster you can really see it sparkle. And through even a small telescope you can behold a countless number of stars.

How to Find It

The easiest way to find 47 Tucanae is to first locate the even larger group of stars, the dwarf galaxy named the Small Magellanic Cloud (SMC). SMC will look like a huge fuzzy patch of gray light to the south of the Toucan constellation among the black background of space; 47 Tucanae will look like a tiny speck of dust blown off the much larger and much more distant galaxy hanging a few degrees to the west of the SMC.

You can also find 47 Tucanae 7–10 degrees below the body of the Toucan constellation when it rides highest in the southern sky. If you are under a dark sky you can make a nearly equilateral triangle with the faint stars of the Toucan; Beta and Delta Tucanae form two corners and 47 Tucanae forms the third.

OCTANS, THE OCTANT

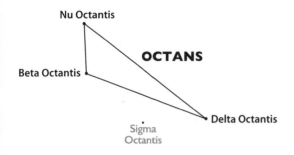

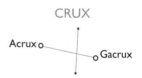

What Is It?
Constellation

Difficulty Level
Difficult

Description

Let's finish up this section with the constellation at the center of it all: Octans. This constellation does not have a long ancient mythology and is barely visible with the naked eye, but if you accept the challenge to locate it, you will be able to find your directions in the southern sky.

Octans was invented by the French astronomer Nicolas Louis de Lacaille during his expedition to South Africa in 1752. Instead of naming this star pattern after characters in Greek mythology (or any mythology for that matter), he named it after a very useful scientific instrument of his day, the octant, which measures angles on Earth and in the heavens.

Despite its dim nature, Octans holds a special place in the sky for Southern Hemisphere observers. The entire sky seems to revolve around it. Occasionally you may see maps of the stars and constellations with grid lines added in. Those are the lines of latitude and longitude of the

Earth projected onto the heavens. In the Southern Hemisphere lots of those lines meet at the South Celestial Pole, which is the point in the sky directly above the South Pole on Earth. As the Earth turns, every constellation in the Southern Hemisphere will appear to spin clockwise around that spot—the spot where Octans happens to be.

The star within the constellation Octans and closest to the South Celestial Pole is called Sigma Octantis. It is a fifth magnitude star, making it visible to the naked eye only from extremely dark skies (and you must have some incredible eyesight). On the other side of the Earth at the North Celestial Pole, 180 degrees from Sigma Octantis, is Polaris, the North Star. Polaris is a second magnitude star and shines 25 times brighter than its southern cousin. But as dim as it is, you can still call Sigma Octantis the "South Star."

How to Find It

The three brightest stars in this star pattern form a skinny right triangle, but are extremely difficult to identify and are so faint you must be away from city lights to even see them. Furthermore, there are two other star patterns that make a triangle shape in the southern sky (Triangulum Australe and Hydrus) and both of them are brighter than Octans.

The good news is that Octans does not move much from night to night or month to month. Every night you can find it in the same place in the southern sky. Whatever your latitude, that is how high in the sky to look for Octans. For instance, if you live at 35 degrees south latitude, Octans will be 35 degrees above the southern horizon.

You can also use the brighter dramatic constellation Crux, the Southern Cross to point you in Octans's direction. Connect the stars on the long part of the Southern Cross and from the fainter Gacrux toward Acrux. Keep going in that direction for another 26 degrees and it will take you to Sigma Octantis and the vicinity of Octans.

The Summer Sky

While the stars and constellations in the southern sky are visible all year round, the rest of them are best seen during certain seasons. Astronomers call these seasonal constellations. The stars and constellations of summer are some of the brightest and easiest to recognize of the whole year, and they provide some of the best stargazing for the beginner.

To view seasonal constellations, you'll be facing mostly east, north, and west. Since only circumpolar constellations (those visible all year) reside in the southern sky, with a few exceptions, we won't need to look much in that direction. Over the course of a night seasonal stars will rise in the east and travel up and to the left until they reach their highest point above the horizon in the northern sky. Then they will start heading down and to the left and toward the western horizon until they set.

Once the New Year dawns, the constellations of summer take center stage and cover the northern quadrant of the sky. To the ancient Greeks this region was filled with mythological creatures interwoven into one large legend. There is one particular entire scene complete with love, love scorned, and death. Would you believe that, in those stars, there is a giant hunter being trampled by a charging bull that has seven women on its back, while nearby two hunting dogs are chasing after a unicorn and a bunny rabbit down by the river? Oh, and there is one added wrinkle: since we're viewing from the Southern Hemisphere, everything is upside down compared to the Greek view.

Learning to find all of these constellations might sound difficult, but we'll go step by step, myth by myth, and explore the placements of the stars and constellations of the summer sky. Along the way we will meet Orion, the Hunter; Taurus, the Bull; Canis Major, the Big Dog; Canis Minor, the Little Dog; Gemini, the Twins; Auriga, the Charioteer; and Eridanus, the River. We'll also take a closer look at the super stars that dwell in these summer constellations. By the end of summer you'll be able to identify them all!

THE SUMMER HEXAGON

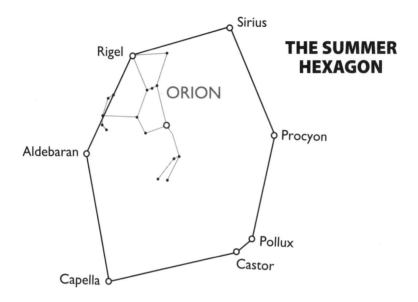

THE SUMMER HEXAGON

What Is It?

Asterism

Difficulty Level

Easy

Description

The nights may be shorter but after it gets dark the summer sky holds the most stellar gems of any season. Eight of the twenty brightest stars in the entire sky—a ring of sparkling multicolored jewels encircling the constellation Orion, the Hunter—shine every summer. These brightest summer stars are often called the Summer Circle or more appropriately the Summer Hexagon. (Of course, from the Northern Hemisphere these stars are visible during their winter months, and so you may see them referred to as the "Winter Hexagon" or "Winter Football" since their shape resembles that of an American football).

At 65 degrees long and 40 degrees wide the Summer Hexagon covers almost half of the entire northern sky. Pointy on two ends, flattened in the two middle halves, the Summer Hexagon is definitely the huge star pattern of the season.

This star chart shows what the sky looks like on a normal summer evening in January and February around 8 or 9 p.m. The Summer Hexagon is outlined and labeled. If you want to picture the actual constellations inside, it would take some imagination. To the ancient Greeks this region was filled with mythological creatures interwoven into one large legend. In this section I will walk you through the entire scene in the summer sky and you can use this asterism as a guide to find many of the stars and constellations throughout the entire season.

How to Find It

To trace out the Summer Hexagon, start by locating Sirius, the Dog Star, the brightest star in the night sky. It can be found at the pointy end of the hexagon that is highest in the northern sky and can even appear straight overhead from some latitudes. Next, going clockwise around the hexagon, you will find Procyon, the Little Dog Star, followed by the Gemini Twins' head stars, Pollux and Castor, which are much lower in the sky toward the northeast. From there, go to bright Capella, which marks the lower point of the hexagon that is opposite Sirius. Make a quick turn up from the horizon and you'll trace the other side of the hexagon toward the Bull's Eye star, Aldebaran. Continue around to Rigel, Orion's left foot, and then head back to Sirius. By doing this, you'll have traced the entire Summer Hexagon! Now that you have your bearings, let's take a closer look at the stars and constellations in and around this huge star formation.

ORION, THE HUNTER

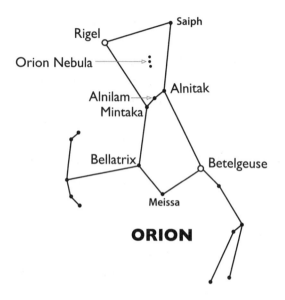

What Is It?
Constellation

Difficulty Level
Easy

Description

Orion is the constellation that conjures the deepest imagination and wonder with just one glance. Something about the placement of the stars ties the entire picture together. Almost every culture in the ancient world associated these stars with a hunter, a giant, or an all-around he-man. His origin story in the Greek myths is murky at best, and no storyteller can seem to agree as to where he came from or how he got to be so tough. Orion typically appears in tales that require a hunter, and in these stories he proves himself to be the best around.

Orion is very easy to see in the night sky and, unlike many constellations, the formation is easy to imagine as a hunter. Unlike the view of these stars from Greece, from the Southern Hemisphere Orion is upside down. It's best to picture him facing you with his right arm (on the bottom right of the constellation) raised high holding a sword and his left arm (on the bottom left of the constellation) bent and holding a shield. But Orion's Belt—made up of the stars named (from right to left) Alnitak ("the girdle"), Alnilam ("the string of pearls"), and Mintaka ("the belt")—ties the whole constellation together. Orion's Belt is a great landmark in the heavens; not only does it identify the giant hunter constellation but it helps point us to other stars and constellations in the summer sky.

Above the belt are two stars that mark Orion's feet (bright blue Rigel and fainter Saiph), and below the belt are his shoulder stars (Bellatrix and Betelgeuse). At the bottom of the constellation is a very faint star marking his head named Meissa. In all of the legends about Orion he is portrayed as being extremely brave, strong, and fearless—but not very bright. And the dim star marking his head displays that appropriately.

How to Find It

To find Orion simply look for three stars of average brightness that form a tight line in the summer skies. Locate them and you have found Orion's Belt. There is no other star pattern visible to the naked eye quite like it.

You can easily identify Orion every evening between December and April. When he rises in the east-northeast in December, he will seem a little unbalanced. He looks like he's tipped over and standing on his head and shoulder. As he reaches his highest point in the northern sky after dark in February, Orion is fully on his head. Standing about halfway up in the northern sky, the outline of Orion's stars is so dramatic that he looks to be perched over the Earth as his bright and noticeable stars shine. As we get to March, when Orion sets in the west-southwest, he once again seems to tip over just before he sets below the horizon. Once April turns to May, Orion is no longer visible in the evening sky. But by spring you can see his trademark belt of three stars in the morning sky, rising in the east just before sunrise.

BETELGEUSE, THE ARMPIT

What Is It?
Star

Difficulty Level
Easy

Description
The most infamous star name in all astronomy is Betelgeuse (because it is most commonly pronounced Beetle-juice and when uttered aloud it makes both kids and adults laugh). Betelgeuse is actually a shortened and edited version of what Arab astronomers called this star: Ibt al-Jauza. Astronomy historians believe that the name originally meant "Armpit of the Central One." After looking at the common illustration of Orion, you would think Betelgeuse would correspond simply with Orion's right shoulder. However, in most depictions of Orion he is raising a club high in the air and thus exposing his armpit.

Betelgeuse is much different than Orion's other bright stars. It is a red supergiant star, and to the naked eye it definitely appears more orange than other stars. When you compare Betelgeuse to Orion's other stars you will see the difference. Bellatrix, Rigel, and the three belt stars are all blue or white. When you look back at Betelgeuse after peering at Orion's other bright stars, you'll agree it is much redder than the others.

Betelgeuse is about 640 light-years from Earth and it is also humongous. If it were our Sun, its volume would stretch beyond the orbit of Jupiter. That means the Earth would orbit inside it! Stars are typically so far away that they only look like little dots in even the biggest telescopes. Betelgeuse, however, is so big that it is one of just a few stars on which astronomers can map features.

But maybe the coolest thing about Betelgeuse is that it will explode. It's so massive that when it dies, it will create a bright supernova. In fact, it should be so bright that you will be able to see it in the daytime. Astronomers do not know when it will go supernova. It could be tomorrow or it could be centuries from now. But keep checking on Betelgeuse, because maybe one night you'll witness Orion's armpit explode!

How to Find It

After you locate the constellation Orion, look for two bright stars below Orion's Belt that could stand in as his shoulders. Since Orion is facing us, look at his right shoulder (on the bottom right of the constellation) and it should be much redder than his left shoulder (on the bottom left of the constellation). That is Betelgeuse and is actually Orion's armpit.

BELLATRIX, THE BEAUTIFUL

What Is It?
Star

Difficulty Level
Easy

Description
Orion's left shoulder, a bright blue-purple star, is named Bellatrix. Bellatrix, which is about 250 light-years from Earth, has one of the deepest blue colors of any bright star. This color tells astronomers that it is much hotter than Betelgeuse. Bellatrix has a surface temperature of about 22,000°C, while Betelgeuse is a relatively cool 3,000°C.

The name Bellatrix comes from ancient Greek and Roman, and it means either "beautiful warrior woman" or "Amazon star." Legend had it that this star rendered all women born beneath it fortunate and talkative. More specifically this star was linked to great Amazon women who were strong, well-spoken, assertive, and tough. Bellatrix has an Arabic name as well. Because it is the first star in Orion to rise above the horizon, it is called Al-Murzim, "the herald of Orion."

Native groups living in the Amazon River basin in South America gave human characteristics to individual stars. They didn't make entire constellations, but they did dwell on the comings and goings of Betelgeuse and Bellatrix. To them each star was a person sitting in a canoe. Bellatrix was a young boy swiftly paddling through the waves with ease. Betelgeuse was an old man struggling with all of his might to keep up.

How to Find It
You can find Bellatrix shining down on you from Orion's left shoulder (on the bottom left of the constellation; remember Orion is upside down). First find Orion's Belt, then look at the two stars marking his shoulders. Betelgeuse is the redder star on your right and Bellatrix is the bluer star on your left.

RIGEL, THE LEFT FOOT

What Is It?
Star

Difficulty Level
Easy

Description
The brightest star in Orion, and the seventh brightest star in the entire sky, is named Rigel. This name can be loosely translated as "left foot of the central one." It twinkles blue-white in color and makes a great contrast to orange Betelgeuse on the other side of the constellation. Rigel figures quite prominently in the ancient Greek legend surrounding Orion's untimely death. Orion bragged that he was such a formidable hunter that he could wipe out all life on Earth. To humble the mighty hunter, the gods thought it would be ironic if Orion was killed by a tiny, almost insignificant creature. He was stung on his left heel by a scorpion—now embodied in the summer constellation Scorpius—and the star Rigel was said to mark the site of the fatal sting.

Rigel is about 860 light-years away but is still one of the brightest stars in the night sky. That means it must be incredibly huge and immensely luminous. Not only does Rigel have a diameter 100 times greater than the Sun but it is also a hot star with a surface temperature of about 11,000°C. This combination of size and brilliance means that Rigel may put out as much as 279,000 times more light than our Sun produces.

Astronomers believe there could be three to five stars in the Rigel system, which means that the star you see with the naked eye is merely the brightest component of a system of suns circling a common center. Rigel is so radiant that it is difficult to see any other stars that may be orbiting it, but astronomers have found direct evidence of at least two other stars near Rigel, and there may be more yet to discover.

How to Find It
Look for blue-white Rigel, the sting on Orion's left foot. From Orion's Belt you will see two stars that could form his feet hanging above. Rigel is the brighter foot star and can be found on his left side of the star pattern.

THE ORION NEBULA

What Is It?
Nebula

Difficulty Level
Moderate

Description
Above the center star of Orion's Belt are three "stars" that form a small sword. The middle of these three appears to be fuzzy in a dark sky. This is not a star, but a star cloud called the Orion Nebula, or M42, a cloud of gas and dust 24 light-years wide that is creating new stars. In this case, this nebula is a huge star factory.

Even though it is about 1,344 light-years away, the Orion Nebula—which has enough material in it to create about 2,000 Suns—is the closest large star-forming region to Earth. Sometimes this nebula can appear fuzzy because you can actually see its nebulosity, the light from the clouds of gas and dust traveling 1,344 light-years to you.

How to Find It
To find the Orion Nebula simply find the three stars that make Orion's Belt. When you look above the belt, you may discover another one, two, or three fainter stars in a line. These three stars make up Orion's sword that is allegedly hanging from his belt. The middle star of the three is the Orion Nebula.

The Orion Nebula is the second brightest object in the sword. If you can only see two stars above Orion's Belt, the brighter one at the top is a star called Nair al-Saif. The Orion Nebula is just half a degree below it.

You can see a little more detail in the nebula by using a pair of binoculars. And through a backyard telescope you can resolve individual stars in the Orion Nebula that are surrounded by a gray cloud of material.

TAURUS, THE BULL

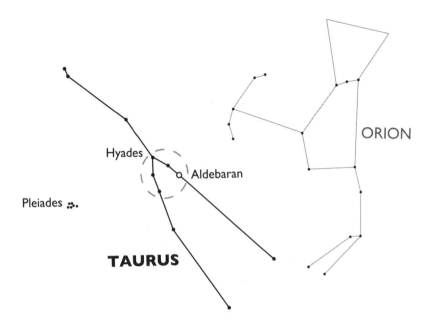

What Is It?
Constellation

Difficulty Level
Easy

Description
According to Greek mythology, Taurus, the Bull was turned into a constellation by the gods in order to protect the Seven Sisters (the Pleiades) from the unwanted advances of Orion. The Seven Sisters got to sit pretty on Taurus's back while Orion fended off the charging beast.

Taurus is one of the most recognizable zodiac constellations in the sky and is arguably the oldest constellation invented. Most astronomers and historians agree that a drawing of the constellation of the Bull is depicted in one of the oldest works of human art found, deep down in a cave near Lascaux, France.

This ancient painting looks suspiciously like the stars in the constellation Taurus, the Bull. Taurus has long horns pointing to the left. The spots in the face correspond to the Hyades star cluster (in the face of Taurus—easily seen with the naked eye). And the seven dots in a tight clump represent the Seven Sisters star cluster. This cave painting is actually a star chart—and it was painted about 17,000 years ago!

Taurus, the Bull lies on the imaginary path in the sky called the zodiac on which the Sun, the Moon, and all the planets appear to move. The zodiac was the earliest form of a calendar, marking the movement of the Sun throughout the year. In ancient days, Taurus held the most important spot—it signaled the return of spring. As a signal of the seasons, the stars of Taurus may have inspired the ancient artist to paint the image of a bull in Lascaux.

How to Find It

Look for an upside-down V shape of five stars to the left of Orion. The V marks Taurus's head. The fainter stars extending out from the V form two long horns.

If you are still having trouble finding Taurus, use the three stars that mark Orion's Belt as pointer stars. Connect the three dots on Orion's Belt and continue this line of sight to the left about 20 degrees (two widths of your fist at arm's length). This will take you to a spot just above a bright, orange-colored star. That star is Aldebaran, or the Bull's Eye, which marks the left side of the V shape of stars.

ALDEBARAN, THE BULL'S EYE

What Is It?
Star

Difficulty Level
Easy

Description
Aldebaran is the brightest star in the constellation Taurus, the Bull. It is super easy to identify, as it stands out as the Bull's Eye, the brightest—and reddest—star in the Bull's V-shaped face. Aldebaran is more than 500 times more luminous than our Sun. If Aldebaran were our Sun, its outer reaches would stretch all the way to the orbit of Mercury.

Although Aldebaran looks to be a part of the larger cluster of stars in the Bull's face, the Hyades, it is not. At only 65 light-years from Earth, Aldebaran is more than twice as close as that cluster. It just happens to be in the same line of sight as those more distant stars.

Despite Aldebaran's sinister, red appearance, this star was a sign of good fortune. More than 5,000 years ago, the ancient Persians designated Aldebaran as one of the Four Royal Stars, the Guardians of the Sky. Each royal star signaled a season of the year, and Aldebaran's proximity to the Sun during the springtime made it the herald of spring. Ancient Hebrews revered Aldebaran as God's Eye. They also called it Aleph or even just A, referring to the first letter of the Hebrew alphabet. The name Aldebaran is an Arabic word loosely translated as "the Follower." What is Aldebaran following? Just continue that line from Orion's Belt past Aldebaran and you will find the most impressive cluster of stars, the Pleiades, or Seven Sisters. As the night goes on, these stars will seem to move from left to right across the sky. This movement gave ancient stargazers the illusion that Aldebaran was following the Seven Sisters.

How to Find It
To find this star, connect the stars of Orion's Belt and keep going to the left. After traveling about 20 degrees, this line will take you just above Aldebaran, which glows with an orange light.

THE HYADES STAR CLUSTER

What Is It?
Star cluster

Difficulty Level
Moderate

Description
If you look carefully at the upside-down V-shaped face of Taurus, the Bull, you may notice that there are several fainter stars scattered around. Most of these stars that you can see with the naked eye are part of an open cluster called the Hyades. Open clusters are dozens, hundreds, or even thousands of stars that were all formed from one massive cosmic cloud. All the stars within a cluster are approximately the same age and distance from Earth. Astronomers have found more than 1,100 open clusters in the Milky Way alone, and they have observed other star clusters in nearby galaxies.

The Hyades is the closest open star cluster to Earth, and it is made up of about 250 stars that all lie about 150 light-years away. Five to ten of the stars can be seen with the naked eye since they are either third or fourth magnitude in brightness. These stars all likely formed from the same nebula, which is a giant cloud of gas and dust.

According to Greek mythology, the Hyades were five daughters of the god Atlas (who is most famously depicted as holding up the world) and half-sisters to the Pleiades. Their appearance in the sky seemed to correspond to the rainy season, and so they were often pictured as crying their eyes out while mourning the death of their brother, Hyas. Their tears became the spring rains.

How to Find It
To find the Hyades, use Orion's Belt as a guide. Connect the three stars in Orion's Belt and continue this line to the left about 22 degrees to the heart of the Hyades, which, along with Aldebaran, fill in Taurus's face. If you extend your left arm and spread your thumb and fingers, then place your thumb on Orion's Belt, your little finger should reach to the Hyades.

THE PLEIADES, OR SEVEN SISTERS, STAR CLUSTER

What Is It?

Star cluster

Difficulty Level

Easy

Description

The Pleiades, or Seven Sisters, is the most famous and impressive naked-eye star cluster in the sky. At first glance the stars in the Pleiades look more like a little cloud, but upon closer examination you may detect five, six, or seven individual stars. However, this cluster is actually a large group of stars moving together in space, about 400 light-years from Earth. Astronomers believe them to be very young and hot stars—formed only 500 million years ago. With binoculars you can see approximately 50 stars in this cluster, and if you carefully scan the boundaries with a telescope you can count all 500.

People with good eyesight think the Seven Sisters resemble a very small dipper, but the ancient Greeks thought the entire group of stars resembled the outline of a dove in the sky. Legend has it that the sisters were trying to flee the amorous advances of Orion and only escaped by divine intervention. The gods took pity on the sisters and turned them into doves. They flew away from Orion and landed up in the sky where they sit today.

That said, the Pleiades are not a constellation of their own. They are part of the larger constellation Taurus, the Bull. When Orion obtained his place in the sky, he continued to harass the Seven Sisters. So the gods established Taurus to reside between the two, forever protecting the sisters from Orion.

How to Find It

To locate the Seven Sisters, use Orion's Belt as a pointer. Draw a line through the three stars in the belt and continue that line to the left. This will take you to the face of Taurus, the Bull, but be sure to keep going. A little more than 10 degrees past Taurus you will run into the Pleiades. It's quite a large hop from Orion's Belt to the Seven Sisters—roughly 35 degrees in the sky—but once you identify them, you'll know why the ancient world was so enamored with these stars.

CANIS MAJOR, THE BIG DOG

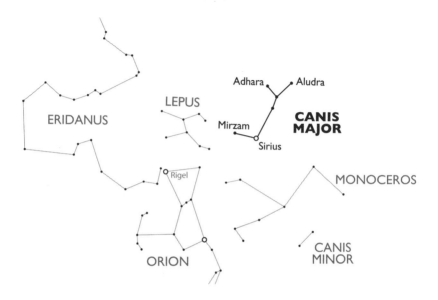

What Is It?
Constellation

Difficulty Level
Easy

Description
The ancient constellation Canis Major ties in with the Greek mythology of Orion and his placement in the sky. According to the legend, after Orion was stung by a scorpion (the winter constellation, Scorpius) and died, he asked the gods if he could bring his two favorite hunting dogs with him into the sky to help fend off the menacing bull, Taurus. The gods agreed to this, and you can find the big dog, Canis Major, and the little dog, Canis Minor, to the right of Orion.

Unfortunately, Canis Major doesn't seem to be helping his master much since he is being distracted by other constellations in the summer sky. He is busy chasing after a unicorn (the constellation Monoceros) and a bunny rabbit (the constellation Lepus) by the banks of the cosmic river (the constellation Eridanus). The unicorn is between Canis Major and Canis Minor. Meanwhile, the rabbit hops below Orion's feet, and the river starts flowing down by Orion's foot star, Rigel.

The outline of Canis Major can be drawn several ways, but it may look more like an upside-down beagle or a Scottie than a vicious hunting dog. The astounding star Sirius can act as an eye with a fainter nose star called Mirzam sticking out to the left. Above Sirius are the dog's neck and two stubby legs (the stars Adhara and Aludra). Canis Major's body gets fainter toward his rear end on the right side of this constellation. But you may be able to detect his back legs in a neighboring constellation called Puppis.

How to Find It

Use Orion's Belt as a guide to find Canis Major. Connect the dots on the three belt stars from left to right. Then extend that line to the right and travel about 20 degrees in the sky. That will take you to the dog's nose and the brightest star in the sky, Sirius.

In November and December you can spot the big dog rising in the east-southeast and it is very high in the northeastern sky by January. In February Canis Major can be so high in the northern sky that it is almost straight overhead after dark. In March and April it is about halfway up in the western sky. You can still see it after dark in May and June lower in the west.

SIRIUS, THE DOG STAR

What Is It?
Star

Difficulty Level
Easy

Description
The brightest star visible from Earth (other than the Sun) is Sirius. The ancients called it the Scorcher, and it lives up to its name by blazing almost twice as bright as the second brightest star, Canopus, which is found in the autumn constellation Carina. As the brightest star in the constellation Canis Major, Sirius has acquired the nickname the Dog Star. Sirius is so dazzlingly bright because it is relatively close to Earth. It is only 8.6 light-years away, making it the seventh closest star beyond the Sun.

Sirius has fascinated people around the globe, and it plays a prominent role in countless cultural myths. Ancient Egyptians worshipped it as the King of Suns and based their calendars on its movements. The rising and setting of Sirius told the Egyptians when to plant, when to harvest, and when the Nile typically flooded. In Hindu mythology Sirius was a hunter and the father of 27 daughters, represented in the 27 daily phases of the moon that were visible throughout its cycle.

How to Find It
Finding Sirius in the sky is about as easy as it gets. Just look for the brightest star in the night sky. Only the planets Venus, Jupiter, and Mars can shine brighter, but there is an easy way to tell them apart from Sirius. Stars twinkle much more than planets, and Sirius often twinkles red, white, and blue when it is low in the sky. If you see a suspiciously bright star noticeably twinkling in the summer sky, you've found the Dog Star, Sirius.

You can also use the stars of Orion's Belt for guidance. Just connect the three stars in Orion's Belt and continue that line of sight up and to the right. After traveling about 20 degrees in that direction, you will run into Sirius.

CANIS MINOR, THE LITTLE DOG

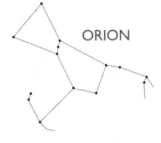

What Is It?
Constellation

Difficulty Level
Easy

Description
Canis Minor is Orion's smaller hunting dog, and it is probably one of the most underwhelming constellations in the entire night sky. With the naked eye you can often only make out two stars in the entire star pattern of Canis Minor: Procyon and Gomeisa.

The gods felt a little sorry about Orion's predicament in the sky. There he was with his belt of three stars just trying to find love with the Seven Sisters. He had fruitlessly chased them on Earth and now continued that pursuit in the heavens. On top of that there was the Bull standing in his way, threatening to trample him every night. He needed help, and so he appealed to the mercy of the gods and was magically granted two hunting dogs to help him with the Bull. Canis Major, the bigger and more helpful of his canine companions, seems up to the task, while Canis Minor seems to be less than an effective assistant. He is tiny compared to the Bull, and with Orion bearing the brunt of the attack, is too far from the action to be of much help. It almost looks like Orion is defending Canis Minor.

How to Find It

To find Canis Minor, first find the three stars in Orion's Belt, then continue that line of sight up and to the right for 20 degrees until you reach Sirius, the Dog Star, in the constellation Canis Major. Then make a 90-degree turn to the right from Sirius and travel another 25 degrees. There you will discover another very bright star called Procyon. This is also called the Little Dog Star and it marks the center of the body of Canis Minor. Just 4 degrees to the right is a fainter, second magnitude star called Gomeisa that stands in for the little dog's head. Congratulations, you have now observed the entire constellation of Canis Minor!

PROCYON, THE LITTLE DOG STAR

What Is It?
Star

Difficulty Level
Easy

Description
Procyon is the eighth brightest star in the sky and shines with a sparkling white light. It is similar in size to Sirius and is also relatively close, residing about 11.5 light-years from Earth.

Other cultures created mythologies around just this star. Babylonians called it Nangar, the carpenter who helped construct the heavens above. Hawaiians used Procyon as a navigational aid for traversing the Pacific on their epic sea voyages. And in Inuit cultures it was called Sikuliarsiujuittuq, the hunter, who, because of his tremendous girth, should not go onto newly formed ice.

Our name, Procyon, comes from the Greek. It means "before the dog," since from Greece it rose above the eastern horizon just before Sirius, the Dog Star, comes up in the southeast. However, as so many things are reversed when viewing from the Southern Hemisphere, Sirius actually precedes Procyon in rising and so you can see Procyon about one month later than Sirius.

Both dog stars, Sirius and Procyon, have a hidden secret. Each of these stars may look like a simple, bright white star, but each has a small companion, called a white dwarf star, revolving around it. These white dwarf stars have nearly the mass of our Sun but are only about 1 percent of its size. That means these companion stars are ultra-dense. These dwarf stars are so small in the sky, you need a large telescope to make them out.

How to Find It
To find Procyon, you just follow the same directions that brought you to Canis Minor. Travel up and to the right from Orion's Belt for 20 degrees until you come to the brightest star in the sky, Sirius. Then take a 90-degree right turn and go 25 more degrees until you reach another impressively bright star. That's Procyon, the body of the little dog in Canis Minor.

GEMINI, THE TWINS

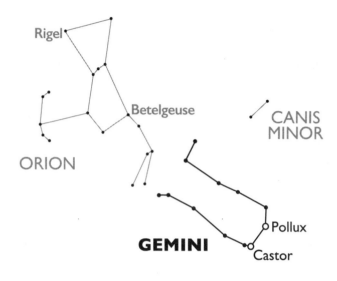

Rigel

Betelgeuse

CANIS MINOR

ORION

GEMINI

Pollux

Castor

Capella

What Is It?
Constellation

Difficulty Level
Easy

Description
The constellation of Gemini, the Twins is one of the oldest and easiest to recognize star patterns in the sky. Two bright stars at the bottom of the constellation named Pollux and Castor stand in for the heads of the twins (Gemini appears upside down in the Southern Hemisphere). However, once you find them and you look a little closer, you will notice that they are not identical twin stars. Pollux is yellow-orange in color and slightly brighter, whereas Castor is blue-white and noticeably dimmer. Pollux is a giant orange star that is about 9 times wider and 43 times more luminous than the Sun. It is situated about 34 light-years away from Earth. Castor, which lies about 51 light-years from Earth, is not just a solitary blue-white star but is in fact a system of six stars that revolve around each other.

The word *Gemini* comes from the Latin meaning "twins." But it was the ancient Greeks who spun the most fanciful tale about these brothers. According to ancient Greek mythology Pollux and Castor shared the same mother, but one was mortal (Castor) and the other was the son of Zeus (Pollux). They grew up developing the greatest bond of friendship. One evening the brothers attended the double wedding of their male cousins (also twins) who were marrying, you guessed it, twin girls! Before the ceremony began, Pollux and Castor accidentally went into the wrong tent, where the twin girls were readying themselves. Well, their eyes met and the twins fell helplessly in love with the twin brides-to-be. The foursome were about to make a quick and romantic getaway from the wedding when the two grooms discovered their plans and stopped them. A terrible fight ensued in which Castor was killed. In Pollux's rage, he killed the two cousins in revenge.

After the brawl Pollux was so saddened that he wished he were dead. He pleaded with the gods to kill him so that he could be with his brother forever in the afterlife. The gods were so moved by Pollux's feelings that they granted his request and immortalized the twins together in the sky to be a sign of fraternal love.

How to Find It

To find Gemini, look for their head stars, Pollux and Castor, using Orion's two brightest stars, Rigel and Betelgeuse, to point the way. Start at Rigel and then draw a line toward Betelgeuse. Keep going and continue that line for about 30 degrees until you find two bright stars, Pollux and Castor. Once you find the head stars of the twins, follow two lines of fainter stars for about 20 degrees (or two times the width of your fist at arm's length) that point upward toward the constellation Orion. With a little imagination you may be able to picture the twins just below Orion's upside-down club.

Gemini is best found on summer evenings, and it rises in the northeastern sky starting in January. By February and March this constellation will only reach one-third to one-half of the way above the northern horizon. During April and May you can catch Gemini lower in the northwest. At the end of May, this constellation says goodbye as it gets lost in the glare of sunset.

POLLUX AND CASTOR

What Is It?
Stars

Difficulty Level
Easy

Description
Pollux and Castor are the brightest stars in the constellation Gemini, the Twins. However, once you find them and look a little closer, you will notice that they are not identical twin stars. One star is yellow-orange in color and slightly brighter—that's Pollux—and the other is blue-white and noticeably dimmer—that's Castor. Pollux is a giant orange star that is about 34 light-years away. It is about 9 times wider and 43 times more luminous than the Sun. Castor lies about 51 light-years from Earth but is more than meets the eye. It is not just a solitary blue-white star but is in fact a system of six stars that revolve around each other. That means if you lived on a planet orbiting Castor, you'd have six suns in your sky!

In African mythology Pollux and Castor were known as the Wise and Foolish Antelopes. Other cultures called them the Two Peacocks, the Two Kids, or the Giant's Eyes. An old Australian myth ties the two bright stars of Gemini in with the star Capella in the constellation Auriga. In that story, Castor and Pollux are two huntsmen named Yurree and Wanjel. They are hunting the elusive kangaroo named Purra (represented by Capella). During the summer when these stars are below the horizon, it is said that the two hunters finally catch the kangaroo and kill it. They then cook his meat over a fire and cause waves of heat to rise above the ground like a shimmering haze.

How to Find It
You can find Pollux and Castor easily by using Orion's two brightest stars, Rigel and Betelgeuse, to point the way. Start at Rigel and then draw a line toward Betelgeuse. Keep going past Betelgeuse and continue that line farther for another 30 degrees and you will quickly be able to identify the only two similarly brilliant stars in that part of the sky: Pollux on the right and Castor on the left.

AURIGA, THE CHARIOTEER

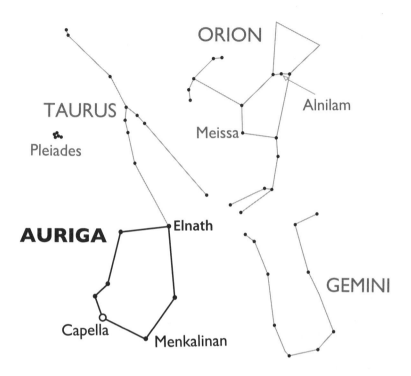

What Is It?
Constellation

Difficulty Level
Moderate

Description
The large summer constellation Auriga features one of the brightest stars in the sky (Capella) and several others that are visible to the naked eye. It was long associated with chariots and charioteers in both Greek and Chinese mythology.

One Greek legend refers to the constellation Auriga as Erichthonius, the son of the gods Vulcan and Minerva. Erichthonius was born deformed and could not walk well. To remedy his situation, he invented the four-horse chariot to get him around the kingdom. He was so respected for his invention that he became the fourth king of Athens. Erichthonius also had a soft spot for crippled or injured animals, his favorite being a little she-goat that is represented by the star Capella. In the sky we are supposed to see Erichthonius holding little Capella and two other goats as they race around the heavens.

It's best to picture Auriga standing on his head just above the northern horizon with Capella on the bottom left. The second brightest star in the constellation is just to the right of Capella and is called Menkalinan, a name that comes from the Arabic words meaning "shoulder of the rein-holder." A star called El Nath, which is also the tip of Taurus's left horn, is Auriga's left heel.

Chinese astronomers saw a similar thing as the Greeks. In Chinese mythology these stars in Auriga were part of the five chariots of the celestial emperors. In other versions many of the stars of Auriga were chairs of the emperors or posts to tie up the horses.

How to Find It

It's not easy to see a guy holding three goats in this star pattern. The constellation itself looks more like a squished pentagon. The surest way to know where Auriga stands is to use the nearby stars in the constellation Orion as guides.

Draw a line downward between Orion's middle belt star, Alnilam, and his dimmer head star, Meissa. If you keep going, after about 30 degrees this line of sight will direct you to the center of Auriga.

Auriga rises in the northeast in December and crawls low across the northern sky. In January and February look for it about a third of the way up in the north. And by March it gets very low in the northwest after dark.

CAPELLA, A NOTORIOUS TWINKLER

What Is It?

Star

Difficulty Level

Easy

Description

Capella is one of the brightest stars in the sky and is similar to our sun in temperature, which means it should look about the same color: yellowish-white. Arab astronomers called this star the Driver, the Singer, and the Guardian of the Pleiades. It was called the Heart of Brahma in India. And in South America this star was one of the favorites among shepherds, who called it Colca. Capella is by far the brightest star in the pentagon-shaped constellation Auriga and was the prized pet of the Charioteer. In Greek its name means "little she-goat."

Capella is a notorious twinkler. Although all stars twinkle, this one seems to attract special attention. When Capella is low in the sky, its light appears to change color, flicker, and dance dramatically. It can even flash alternately red, white, and blue. The effect is especially noticeable on October evenings when Capella is rising in the northeast.

How to Find It

Capella and Sirius are the stars at the pointy ends of the Summer Hexagon and are also the brightest stars in this expansive star pattern. While Sirius remains very high above the horizon, Capella can travel quite low in the heavens. During the heart of summer it will never get very high above the northern horizon. The higher it appears, the less it will twinkle, but you can't mistake its dazzling appearance each summer evening.

The guide stars from Orion that brought you to Auriga can also take you to Capella. Start with the middle star in Orion's Belt, Alnilam, and draw a line from there through Orion's head star, Meissa. Continue that line of sight past Meissa and travel about 36 degrees. That will take you right to Capella.

ERIDANUS, THE RIVER IN THE SKY

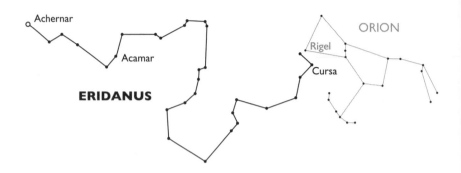

What Is It?
Constellation

Difficulty Level
Moderate

Description
Eridanus is an incredibly long and curvy constellation that, from beginning to end, stretches more than a third of the way across the sky. It is best seen in its entirety during January and February, but the southern portions of the constellation can be seen any evening between September and March.

The brightest star in Eridanus is named Achernar, which comes from Arabic, meaning "the end of the river." But Achernar can be confused with another star farther north in Eridanus named Acamar. Astronomers like the ancient Greeks and Arabs living in the Northern Hemisphere could only see so far south along the river constellation. Achernar was too far south for most northern civilizations to see. But the star Acamar was observable, so for some that star became the end of the river. As explorers traveled farther south across the globe, they discovered that there was more to the river. They were treated to an even brighter end of the river, the star Achernar, and it became the official river's end.

The name Eridanus may originate from "the Star of Eridu," an ancient Babylonian constellation that likened it to the source of underground fresh water. Since the length of Eridanus traverses the northern and southern sky it was also likened to the Nile River, which flows in roughly similar directions. Indian mythology also refers to these stars as a river, with some groups equating it to the river Ganges.

This constellation is also the home of the so-called Eridanus Supervoid—a slightly colder region of space that seems to be lacking in galaxies. When astronomers look in almost every direction they find galaxies upon galaxies littering the universe. For some reason, a section of Eridanus is missing them. One of the wilder and unproven theories is that this is where a parallel universe may be butting up against our universe. It is highly unlikely, but still fun to imagine Eridanus as a river to another universe!

How to Find It

Eridanus has many fainter stars within its boundaries and may be difficult to see in its entirety from urban locations. That said, the best way to identify sinuous Eridanus is to locate the beginning and end of the river. Start at the source of the river by finding the constellation Orion with his belt of three stars halfway up in the northern sky. The brightest star in Orion is above the belt and slightly to the left and is named Rigel. Eridanus begins at the semi-bright star named Cursa that is off to the side and below Rigel. From Cursa you travel higher in the sky following a trail of fainter stars to the zenith point (or straight overhead). Turn around and then come down the other side of the sky toward the southern horizon. After going another 15 degrees past the zenith you'll find semi-bright Acamar. That's not the end of the river, however, so keep going down and after another 20 degrees, the bright star Achernar will signal the river's end.

ACHERNAR, THE END OF THE RIVER

What Is It?
Star

Difficulty Level
Easy

Description
Achernar is a very bright star 140 light-years from Earth that marks the southern terminus of the river constellation Eridanus. It is the tenth brightest star in the entire sky and shines with a bright blue color. Achernar's color is an indication of its temperature. The bluish hue tells you that it has a scorching surface temperature of about 15,000°C and is a much hotter star than our 5,500°C yellow Sun. Achernar is not only hot, but it is huge. It is seven times more massive than the Sun and shines over 3,000 times brighter.

Achernar is also the most squished star that we know of in the galaxy. While most stars that you see are nearly perfect spheres, Achernar is an oblate spheroid (a flattened sphere like a piece of M&M's candy). Despite its massive size, this star is spinning so rapidly that its gases have migrated to its equator and given it a significant bulge.

How to Find It
Achernar is always the southernmost star in the constellation Eridanus, the River. You may first notice it as the only bright star in the southern sky since the other bright stars in the Southern Cross and Centaurus are about 60 degrees distant. Achernar is best seen during summer evenings when it sits high in the southern sky. However, it is so far south that it is nearly circumpolar (meaning it is almost visible all year round at some point of the night).

The other way to locate Achernar is to start at the source of the river constellation Eridanus. Trace the curvy path of fainter stars from the star Rigel in Orion in the northern sky, through the zenith, and part of the way down the southern sky. There Achernar will shine like blue waters at the mouth of the river.

The Autumn Sky

Each change of season provides an entire flock of new stars and constellations to learn. As March and April arrive, the stars and constellations of summer such as Orion, Canis Major, and Eridanus are still visible, but instead of being in the eastern half of the sky they are now positioned more toward the west. A new bevy of autumn constellations has arrived to take their place and inhabit the eastern sky.

The autumn sky is definitely not as dramatic and star-studded as the summer sky. While you have the Summer Hexagon and so many bright stars visible in the summer sky, there are only three first magnitude stars among all the autumn constellations: Regulus, Arcturus, and Spica. Together they make a gigantic star pattern called the Autumn Triangle. This is not an official constellation, but the Autumn Triangle is a noticeable asterism that spans three constellations (Leo, Boötes, and Virgo) and makes a terrific jumping-off point to the other fainter and subtler constellations of autumn, including a large southern constellation full of celestial treasures called Carina.

In this section we will take a closer look at the three constellations that make up the Autumn Triangle—Leo, the Lion; Boötes, the Bear Driver; and Virgo, the Maiden—as well as the major stars that lie within these three constellations. Next we will explore the large constellation Carina with its many notable stars and clusters. Then we will check out the largest constellation in the entire sky, Hydra, the Many-Headed Snake, which stretches almost across the entire northern sky during the heart of autumn. We'll continue our tour of the autumn sky by identifying several small and subtle star patterns in Coma Berenices, Berenice's Hair; Corona Borealis, the Northern Crown; and Cancer, the Crab.

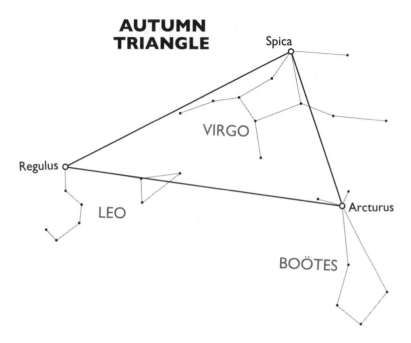

What Is It?
Asterism

Difficulty Level
Easy

Description
If you connect the dots from the three brightest stars in the autumn sky, you can form a giant triangle. Known as the Autumn Triangle, this unofficial constellation, or asterism, is 60 degrees long and 30 degrees wide, and it covers most of the northeastern sky during early autumn and much of the northwestern sky in late autumn. Note: what I call the "Autumn Triangle" in this chapter is an informal star pattern regularly called the "Spring Triangle" in the Northern Hemisphere, where the seasons are reversed. The Autumn Triangle is huge and is super easy to identify since these are the only bright stars in this section of the sky.

The three stars that make up the corners of the Autumn Triangle are named Arcturus, Spica, and Regulus. These stars can help you identify some rather indistinct star patterns around the autumn sky. Arcturus is the brightest star in the cone-shaped constellation Boötes, the Bear Driver; the star Spica can show you where the hand of Virgo, the Maiden resides; while Regulus shows you the heart of Leo, the Lion.

How to Find It

The easiest way to find the Autumn Triangle is to look for the stars at each of the triangle's three corners: Regulus, Spica, and Arcturus. There are no other bright stars equal to them in the autumn sky, so you can easily identify them by their position within the giant star pattern. Regulus leads the way in this asterism and is the first star of the Autumn Triangle to rise and the first one to set. You can start to see it shining its blue-white light in the eastern sky after dark in March. It and the constellation Leo is your signal that autumn is almost here.

Spica is the next to pop up above the eastern horizon along with the constellation Virgo, followed closely by Arcturus and its constellation, Boötes. Arcturus will be farther to the left (or closer to the northeast), while Spica will be over to the right (or closer to due east). This is a giant triangle covering most of the northeastern sky as autumn breaks. The angular distance from Regulus to Spica is 54 degrees. Regulus to Arcturus is the longest side of the Autumn Triangle at 59 degrees. And even the closest corner stars, Spica and Arcturus, are still 33 degrees apart. You can measure these angles with your fist at arm's length—remember each fist width is equal to 10 degrees. So you can see that the Autumn Triangle is huge!

You can start to see the entire Autumn Triangle after dark in April. But the best months to see the complete asterism are during May and June when it stretches high across the northern sky. By July, Regulus sets in the west and only the other two stars in the Autumn Triangle remain. Spica and Arcturus can linger into the beginning of September, outlasting the autumn season.

Sometimes the planets wander through to mess up the outline of the Autumn Triangle. Both Regulus and Spica lie near the ecliptic—the pathway of the planets. So do not be surprised to find extra bright lights between Regulus and Spica from time to time. These planets may ruin the perfect triangle shape, but they make for great bonus objects to observe.

LEO, THE LION

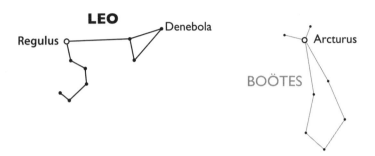

What Is It?
Constellation

Difficulty Level
Easy

Description
There is no surer sign in the heavens that autumn is back than seeing the constellation Leo, the Lion in the sky. Leo is recognizable by the six stars shaped like a backward question mark—also called the sickle—which form its head. The bright star Regulus is the dot in the question mark and designates this King of the Beasts. The back end of Leo is marked by a triangle of stars, the farthest east being his tail, Denebola.

Ancient Greek mythology equated the constellation Leo with the Nemean Lion, whose hide was so tough that no sword or arrow could pierce it. This lion was menacing the countryside, killing villagers left and right. Hercules (also a winter constellation), the demigod and all-around he-man, was called in to take care of the situation as the first of his twelve labors (the twelve "impossible" jobs he was forced to accomplish in order to atone for killing his wife and kids during a fit of madness).

Instead of attacking Leo with a sword, Hercules wrapped his muscled arms around the beast's neck and strangled the lion to death. If you ever see pictures of Hercules later in his life, he's always wearing a lion skin. That's Leo's hide. It was like wearing bulletproof armor, and it protected

Hercules from attacks. One question: How did Hercules cut Leo's hide to fashion his outfit if it was impenetrable? Answer: He used the lion's own claws to do the job.

How to Find It

Leo is an easily recognized star pattern because of the lion's distinct head and rear end. Look for a group of stars that form a sickle shape or backward question mark. But now remember that the view from the Southern Hemisphere is almost upside down compared to the view from ancient Greece. This means Leo flies across the northern sky upside down. Leo's brightest star, Regulus, stands in as the dot at the end of the handle of the sickle shape of stars that is supposed to be the lion's head and mane.

In March look for Leo low in the east after dark. His head rises first and lies to the right of the Summer Hexagon (the incredibly bright ring of stars surrounding the constellation Orion). On April evenings you can spy the stars of the lion halfway up in the northeastern sky. In late autumn he's roaring high in the northern sky, and by the start of winter Leo is over in the northwestern sky.

You can also use two really bright stars to help triangulate the lion's position. Connect a very long 100-degree line between Sirius, the brightest star in the sky, and Arcturus, the brightest star in the constellation Boötes. Leo's brightest star, Regulus, will reside halfway between the two.

REGULUS, THE LITTLE KING

What Is It?
Star

Difficulty Level
Easy

Description
If Leo, the Lion is the king of the beasts, then he needs a royal star to go with him. And indeed, Regulus delivers. The name comes from Latin and means "little king." Regulus was one of the Four Royal Stars of ancient Persia, along with Fomalhaut (in Piscis Australis), Aldebaran (in Taurus), and Antares (in Scorpius). Each royal star reigned over a season of the year, and Regulus was the star whose departure from the sky signaled the changing seasons.

Regulus is blue-white in color and lies about 78 light-years from Earth. Astronomers have discovered that the light you see from Regulus actually comes from four stars that revolve around one another. The largest one, called Regulus A, is almost four times more massive than our Sun. Regulus A also spins rapidly, causing its gases to accumulate more around its equator. Astronomers characterize Regulus's shape as an oblate spheroid. You can liken it to a flattened egg. This is the star you are seeing in the spring sky with your naked eye, while the other three stars in the system require huge telescopes to spot.

How to Find It
Leo's head stars look like a sickle shape or backward question mark. Regulus can be found at the handle of the sickle or, if you prefer, the bright dot at the end of the backward question mark. Look for this formation of stars in the northeastern sky in early autumn, high overhead in late autumn, and over in the northwestern sky as winter begins. Regulus stands out as the brightest star in the constellation and in its quadrant of the sky.

BOÖTES, THE BEAR DRIVER

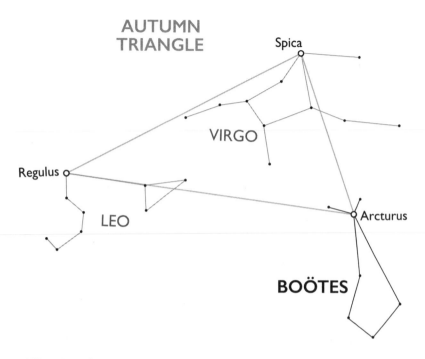

What Is It?

Constellation

Difficulty Level

Moderate

Description

Boötes (pronounced Bo-OH-teaze) is a kite-shaped constellation that first appears in the evening skies of April looking like a wide tie hanging from the invisible neck of an invisible businessman.

In the Northern Hemisphere, Boötes is known as the Bear Driver, since he can be found chasing after the big bear constellation, Ursa Major. From the midlatitudes of the Southern Hemisphere, however, Ursa Major is either too low in the northern sky or below the horizon at all times and can't be seen.

So let's look to another ancient Greek myth that also deals with Boötes. In this story, this constellation represents Icarius, an immigrant farmer who moved to Athens to start growing grapes. As a welcoming gesture he invited the god of the grape harvest, Dionysus, over for a feast. The god was so touched that he imparted a divine secret to Icarius: he taught him the art of wine-making.

Icarius tried it out and made a small batch for his neighbors. It quickly turned into a great party and a good time was had by all. But the next day when the villagers awoke with severe hangovers and felt so sick they naturally assumed that Icarius had poisoned them. The neighbors tracked Icarius down and killed him. Dionysus felt so bad about the turn of events that he honored Icarius by turning him into the constellation we call Boötes today.

How to Find It

From the Northern Hemisphere and Tropics, the handle of the Big Dipper can point you toward the stars of Boötes. However, from latitudes farther south than 25 degrees the Big Dipper is not visible above the northern horizon, so you will need a new trick to finding this kite-shaped constellation. Face north and locate the expansive Autumn Triangle of three bright stars: Arcturus, Regulus, and Spica. They will be the only bright stars in the entire northern sky during the evening hours this season. Arcturus is by far the brightest of the three stars and is more than twice as brilliant as Regulus and Spica. Once you locate Arcturus, see if you can make out the kite-shaped pattern of stars hanging off of this dazzling star.

Of the three constellations in the Autumn Triangle, Boötes is the farthest north and is visible for the shortest amount of time. It rises in late April and early May in the northeastern sky and stays relatively low in the north as those autumn nights go on. In June and July it is about halfway up in the northern sky. By August Boötes is low in the northwest after dark and by September it is pretty much lost in the glare of the sunset for the year.

ARCTURUS, THE BRIGHT

What Is It?
Star

Difficulty Level
Easy

Description
The name Arcturus comes from Latin and means "Guardian of the Bear"; for ancient Greeks Arcturus was visible relatively close to the northern constellation Ursa Major, the Big Bear. Polynesian astronomers called this star Hōkūleʻa, meaning the "star of joy," and used it as a guide star to help them navigate around the Pacific Ocean. Since Arcturus can appear straight overhead from the Hawaiian Islands, Polynesian sailors used its elevation to determine their latitude. When Arcturus was overhead, they sailed east or west until they ran into the islands.

As the fourth brightest star in the entire sky, Arcturus's intense brightness will definitely attract your attention. When you first see it, you may wonder if it is a plane or UFO. Only Sirius (visible in the summer sky), Canopus (in the autumn sky), and Alpha Centauri (up in the winter sky) are brighter.

Even with the naked eye you will notice that Arcturus shines with a decidedly orange tint. This color tells astronomers that Arcturus has a surface temperature of about 4,000°C (a much cooler star than our 5,500°C yellow Sun). Arcturus is about 36.7 light-years from Earth, making it one of the closer stars. It is an orange giant star with a diameter 25 times that of the Sun, and it shines 170 times brighter.

The vivid light of Arcturus famously triggered the start of the 1933 World's Fair in the US city of Chicago. When this star's light fell on a sensor, it activated a switch that illuminated the fairgrounds on opening night.

How to Find It
To find Arcturus, look to the northern sky during the autumn months to identify the Autumn Triangle, the large asterism that spreads across nearly 60 degrees of the sky. Arcturus is the brightest star in the Autumn Triangle and also appears closest to the northern horizon.

VIRGO, THE MAIDEN

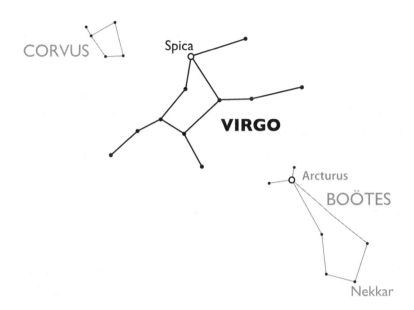

What Is It?
Constellation

Difficulty Level
Moderate

Description
Virgo is known as the virgin or maiden. It is a huge constellation (only Hydra is larger) that appears in the southeastern sky on April evenings. Illustrations of the virgin in the sky show her reclining serenely while holding an ear of wheat in one hand and a dove of peace in the other. However, it is difficult to imagine Virgo's silhouette when looking at the night sky. Many of her stars are dim, and the figure really doesn't look much like a maiden, but she is still moderately easy to find if you can identify her brightest star, Spica (pronounced SPY-kah).

The ancient Greeks likened Virgo to the daughter of their chief god, Zeus, and Themis, the goddess of justice. Virgo lies in the sky next to another zodiac sign, Libra, the Scales. This location displays her association with the scales of justice. Unfortunately, Virgo is not holding the scales since the constellation Libra lies at the maiden's feet.

In ancient Egyptian mythology Virgo was associated with Isis, the supreme mother goddess. As legend has it, one day Isis was eating corn in the sky when a monster of enormous size and viciousness named Typhon came upon her and chased her far and wide. As Isis fled from the great beast, she dropped pieces of corn all across her path. These kernels of corn turned into stars and became the Milky Way.

In India, Hindu astronomers believed Virgo to be Kanya, the maiden and mother of the great Krishna. Arab astronomers initially included Virgo in a giant Lion constellation that covered a huge swath of the spring sky, while others in the Middle East called the stars of Virgo "the Barking Dogs." Over time these astronomers changed this constellation to conform to the Greek myths and, toward the end of the first millennium A.D., began to call Virgo by the name Al Adhra al Nathifah, which translates to the Innocent Maiden.

How to Find It

You can locate Virgo by finding Spica's place among the stars of the Autumn Triangle. It is the second brightest one in this asterism (Arcturus is significantly brighter). You can also use Arcturus and the outline of the constellation Boötes to point you toward Virgo. Connect a line between the star at the bottom point of Boötes called Nekkar and Arcturus. Keep going another 30 degrees and you will run into Spica and the body of Virgo.

Virgo rises in the east after sunset in April. In May she is higher in the northeast and can climb to about two-thirds of the way up in the northern sky on June evenings. During July and August Virgo is about halfway up in the northwest and you get the last glimpses of her in September just above the western horizon.

SPICA, THE BLUE-WHITE DIAMOND

What Is It?
Star

Difficulty Level
Easy

Description
Spica is the real star of Virgo. It is a brilliant blue-white in color and lies 260 light-years away. It has been called the Queen Star of the season and the Star of Prosperity. However, in Latin, Spica literally means "ear of wheat," which Virgo is typically depicted holding in her left hand.

The Egyptians had a special affinity for Spica because of its location along the zodiac, and they built many temples to its movement. At key times in the year, the light of Spica would penetrate through deep shafts into these temples.

Spica lies near the celestial equator, which means it is prominently visible every night for about six months (March–September). Its rising at sunset marked the official beginning of the spring planting season for cultures across the Northern Hemisphere. The setting of Spica at sunset signaled the autumn harvest time.

Spica is also near the ecliptic—the apparent pathway that the Sun, Moon, and planets wander. About once a month, the Moon cozies up to Spica and can even cover it up (this is called an occultation). Planets are also frequent visitors to Virgo and Spica, and it is not unusual to see Venus or Mars nearby.

How to Find It
Spica is the brightest star in Virgo and the only one in the constellation that stands out. First locate the three stars in the Autumn Triangle. Arcturus is the brightest of the three, with Spica coming in as second brightest and lying about 30 degrees above and to the left of Arcturus for most of the autumn. Arcturus, along with a trapezoid shape of four semi-bright stars that makes the constellation Corvus, the Crow, bracket Virgo. During the heart of autumn you can find Arcturus well below Spica, with Corvus just above it as they appear to move across the sky.

CARINA, THE KEEL OF THE SHIP

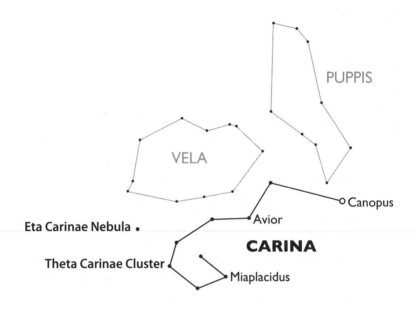

What Is It?
Constellation

Difficulty Level
Easy

Description

Carina is an expansive constellation 20 degrees wide and 35 degrees long that sails high across the southern sky during late summer and early autumn. However, it used to be bigger than what you can see today. The constellation was best detailed in A.D. 150 by the ancient Greek astronomer Ptolemy. He included the stars of Carina along with many others to its north to create one great ship that he called Argo Navis.

The legend began when a king, with a son named Phrixus and a daughter named Helle, left his wife and remarried an evil woman. The new wife was jealous of the children and planned to sacrifice them to the gods. At the last moment their biological mother sent a magical, winged, golden ram to rescue Phrixus and Helle to fly them away to the east. Unfortunately, the daughter fell off the flying ram to her death in an area still called the Hellespont (named in her honor). Phrixus landed safely, sacrificed the ram to Zeus, and gave the Golden Fleece to another king named Aeetes (whose daughter Phrixus later married).

Years later a Greek hero named Jason gathered a group of explorers called the Argonauts, who set sail (on the ship called the *Argo*) to reclaim the Golden Fleece. The Argonauts, which included Greek heroes Hercules, Orpheus, Castor, and Pollux, among others, had a series of epic adventures and ultimately succeeded in their quest.

The *Argo* inspired the star pattern Argo Navis and remained a giant constellation until the eighteenth century. But it was so expansive that astronomers decided to break it up into three parts: Carina became the keel of the ship while the nearby constellation Puppis became the poop deck and the constellation Vela stood in for the sails.

Carina includes the second brightest star in the entire sky, Canopus, as well as two other first magnitude stars named Miaplacidus and Avior. And peering much deeper into the space occupied by Carina you can discover more subtle gems of the southern sky like the Eta Carinae Nebula and the Theta Carinae, a star cluster similar to the Pleiades or Seven Sisters star cluster, which is visible in the northern sky.

How to Find It

Canopus, the second brightest star in the entire sky, is all you need to identify the constellation Carina. Although Carina stands highest in the sky during the early spring evenings, you can still spy Canopus rising in the southeastern sky even in January. The constellation will look like a dangling fishhook with dazzlingly bright Canopus at the top and Theta Carinae and Miaplacidus at the bottom. But once autumn begins, Carina will stand very high in the south, approximately 60 degrees above the horizon. Canopus will be at the far right of the ship-like star pattern 60 degrees above the southwestern horizon. You can also look for the fainter outlines of the constellations Puppis and Vela that will be above the curve of stars forming Carina and will appear almost directly overhead.

By April and May the constellation will start to appear lower in the southwestern sky with Canopus leading the way. Canopus will be low in the southwestern sky in May and start to set on late June nights. The remainder of the constellation will set during July and August. Between September and December, the constellation is mostly below the southern horizon and not visible from mid-southern latitudes.

CANOPUS, THE PILOT

What Is It?
Star

Difficulty Level
Easy

Description
Canopus is easily the brightest star in the constellation Carina and the second brightest star in the night sky (only Sirius, the Dog Star in the constellation Canis Major is more brilliant). While stargazers in most of the United States, Canada, Europe, and Asia can only see Sirius, observers in the Tropics and most of the Southern Hemisphere can see both brilliant stars. Canopus appears to be stark white when higher in the sky and resides about 310 light-years from Earth. Its name comes from a minor character named Canopus who piloted a ship in the ancient Greek myth about the Trojan War.

Canopus served as a guide star for Bedouins and other groups when crossing the deserts or Africa and the Middle East. Polynesians used it to help navigate the South Pacific and associated it with a larger bird constellation called Manu that stretched from Canopus to Sirius and another bright star named Procyon. In New Zealand, the Maori people had several names for Canopus but usually highlighted its special and solitary place in the southern sky, calling it Stand Alone.

How to Find It
From January to May you can watch the two brightest stars in the sky move together each night. Face south and you will see a dazzlingly bright star. That is Canopus. But if you scan the sky further to the north, about 35 degrees away you will see an even brighter star. That is Sirius.

Canopus is the also the first major star in the constellation Carina to clear the southeastern horizon, so you can identify it even in November and December low above the horizon where it will twinkle wildly. By January and February it will be halfway up in the southeast and by March and April will be about 60 degrees up in the south. During May and June evenings it will get lower in the southwest and by July, Canopus will set with the Sun.

ETA CARINAE NEBULA

What Is It?
Nebula

Difficulty Level
Moderate

Description
The Eta Carinae Nebula is a humongous region filled with stars under construction, newly formed star clusters, and one stellar behemoth nearing the end of its life. This strange combination of unborn, newly born, and dying stars 7,500 light-years away is visible from the Tropics and the Southern Hemisphere.

The most famous star inside the Eta Carinae Nebula is simply called Eta Carinae. This is a star about 4 million times brighter than our Sun. If it were a lot closer than 7,500 light-years away, Eta Carinae would be one the brightest stars in the sky. And, at one point, it actually was. For four days in 1843, Eta Carinae erupted and brightened to the point that it was equal to Alpha Centauri, the third brightest star in the sky.

Eta Carinae is big and bright, but the Eta Carinae Nebula hosts an even larger resident star. WR 25 is a rare type of massive object (called a Wolf-Rayet star) that is the most luminous known star in the galaxy. It shines 6.3 million times brighter than our Sun. Unfortunately, WR 25 cannot be seen with the naked eye. It is surrounded by thick envelopes of gas that make it appear dimmer than it is.

How to Find It
The Eta Carinae Nebula will look like a small, cloudy star at first. But with a pair of binoculars you can make out more detail in the nebula. It can be found about 15 degrees to the right of the Southern Cross. If you connect a line through Mimosa (Crux's second brightest star) and Epsilon Crucis (Crux's fifth brightest star) and continue that line for another 12 degrees you will run into the Eta Carinae Nebula.

THETA CARINAE CLUSTER

What Is It?
Star cluster

Difficulty Level
Easy

Description
This star cluster has a lot of nicknames. It is called the Theta Carinae Cluster (after the brightest star in the cluster) and IC 2602 (IC stands for Index Catalog, a group of deep space objects charted in the late-nineteenth and early-twentieth centuries), but its popular name is the Southern Pleiades since it bears a resemblance to a famous star cluster visible in the Northern Hemisphere called the Pleiades. At about 479 light-years from Earth, the sixty stars in this open cluster are similar in distance to their northern cousins. But they are a much fainter gathering of stars, and although you can easily find them with the naked eye, they look best when viewed through a pair of binoculars.

Theta Carinae is an open cluster, meaning all the stars were formed from one giant cloud of gas and dust and are linked gravitationally, but are still far enough apart to see "open" space between the individual stars. The cluster is so expansive that it takes up nearly 1 degree of the sky (twice the width of the Full Moon).

How to Find It
The stars in the Theta Carinae Cluster are so far south that they are actually circumpolar from mid-southern latitudes and are thus visible almost all year round in the southern sky. The easiest way to find them is to start at the Southern Cross, which lies nearby. Connect the dots on the two faintest stars in the cross, Gamma and Delta Crucis. Travel from brighter Gamma to dimmer Delta and continue that line of sight for another 12 degrees. You will run right into the brightest star in the cluster, Theta Carinae, and be able to see the rest of the cluster of stars surrounding it. At first the cluster may look like a little cloud to the naked eye, but relax your eyes and you should be able to make out individual stars in the cluster.

THE FALSE CROSS

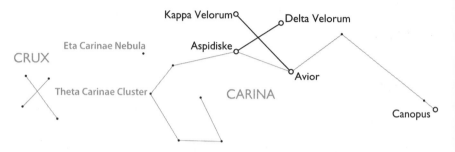

FALSE CROSS

Kappa Velorum
Delta Velorum
Eta Carinae Nebula
Aspidiske
CRUX
Avior
Theta Carinae Cluster
CARINA
Canopus

What Is It?
Asterism

Difficulty Level
Easy

Description
The Southern Cross is the easiest, most-recognizable star pattern in the southern sky. However, it can be confused with another larger yet fainter star pattern often called the False Cross that straddles the constellations Carina and Vela. The False Cross is not an official constellation and is therefore an asterism. It comprises Aspidiske and Avior (two stars from Carina) that make up the left side and bottom of the cross and Kappa Velorum and Delta Velorum (from Vela) that make up the top and right side of the cross. Remember that in the ancient days Carina and Vela were linked into one larger constellation of Argo Navis and so it is strange that astronomers broke up the ship and the False Cross.

How to Find It
How do you tell the difference between the real Southern Cross and the False Cross? The Southern Cross is only 4 degrees wide and 6 degrees long while the False Cross is about 6 degrees wide and 9 degrees long. The False Cross seems like a fainter and less-impressive copy of the original.

The confusion can arise because the False Cross rises before the Southern Cross. As all of the objects in the sky seem to move from east to west, the False Cross is further west and will seem to lead the way. Some people might first see the False Cross and incorrectly say, "Ah, there is the Southern Cross," and an hour later when the smaller and brighter true Southern Cross becomes visible, they will see their error.

Look for the False Cross about halfway along the length of the constellation Carina. Its center is about 20 degrees from brilliant Canopus on one side and about 25 degrees from the center of the Southern Cross.

COMA BERENICES, BERENICE'S HAIR

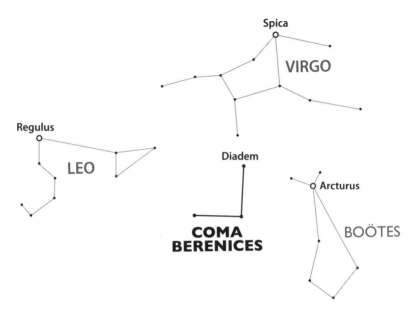

What Is It?

Constellation

Difficulty Level

Difficult

Description

The faint smattering of stars in the autumn sky called Coma Berenices has a hair-raising tale. In Greek mythology Berenice was the beautiful queen of Egypt known for her flowing tresses. When her husband went off to war, Berenice asked Aphrodite to protect her beloved in battle. If he returned to her safely, she would cut off her long hair as a gift to the goddess. When the king returned unharmed to her side, Berenice stayed true to her word and lopped off her hair. The hair was placed in the temple where it mysteriously disappeared. Who dared to take the queen's beautiful hair?

Heads were going to roll if the culprit was found. Luckily a court astronomer came to the rescue—he found the missing locks up in the heavens. The hair was such a pleasing sacrifice to Aphrodite that she took it and placed it in the sky for all to see. And so the glory of Berenice's hair reached new heights and lives on in the stars.

One of Coma Berenice's brightest stars is called Diadem. But Diadem is no dazzling jewel of the night sky and is challenging to find since it is so dim compared to the bright stars in the Autumn Triangle. Even so, with a lot of imagination maybe you can envision this star as a jewel in the Queen's crown or merely another lock of flowing hair streaming out of this nebulous constellation.

How to Find It

Coma Berenices is incredibly faint and can only be detected from a dark sky. On some star charts astronomers outline the three brightest stars in the constellation with a right angle shape, but this doesn't capture the true nature of the entire star pattern. At first glance it will look like a diffuse cloud about 20 degrees below Virgo and about halfway between the constellations Boötes and Leo. When you squint in this region of space, you may be able to make out a swarm of individual stars that evoke an image of flowing hair.

HYDRA, THE MANY-HEADED SNAKE

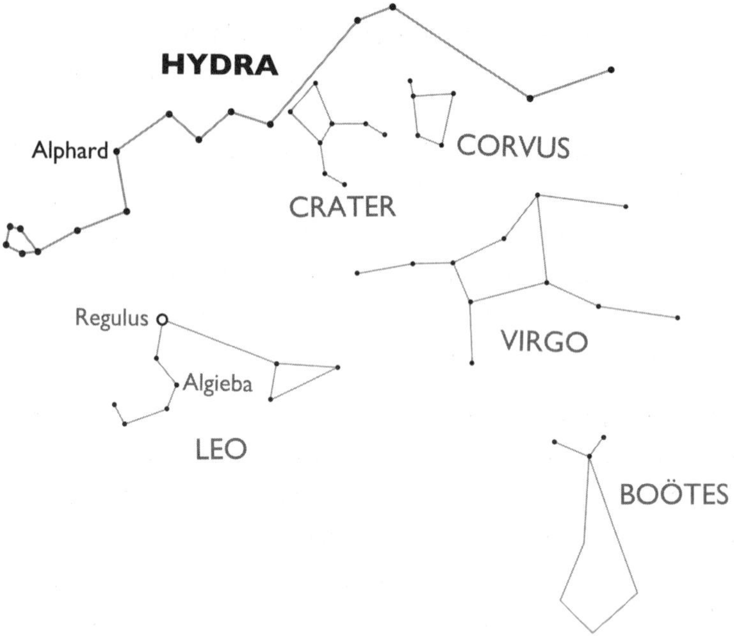

HYDRA

Alphard

CORVUS

CRATER

Regulus

Algieba

VIRGO

LEO

BOÖTES

What Is It?
Constellation

Difficulty Level
Difficult

Description
Hydra is the largest constellation and coils across most of the northern sky every April and May evening. From heads to tail Hydra spans more than 100 degrees. Remember that your fist at arm's length equals about 10 degrees. So Hydra is ten fist widths long—which is more than one-quarter of the way around the horizon!

Hydra also has company. Sitting below his back are two fainter constellations called Corvus and Crater. Corvus is definitely the brighter of the two and is a crafty crow, while Crater is a cup of water.

Like the constellation Leo, Hydra was a victim of Hercules (also a winter constellation), the Greek warrior and he-man. After strangling the Nemean Lion, Hercules had to kill the seven-headed Hydra of Lerna. This was the second of his twelve labors, and it was not an easy one to accomplish. The Greeks believed that Hydra was a terrible monster with seven, eight, or even nine heads. People had tried to kill Hydra before but discovered that every time they cut off one of the heads, two more would grow back in its place!

Hercules went to slay the monster with help from his cousin Iolaus. Whenever Hercules cut off a head, he would signal Iolaus to come over with a hot iron and cauterize the wound shut so that no heads could grow back. It worked! After the seventh head fell, the Hydra was dead and joined Leo, the Lion in Hercules's monstrous pet cemetery in the sky.

How to Find It

To find Hydra, first find the constellation Leo, the Lion. Although typically lower in the sky than Hydra, Leo's sickle shape of stars is much more distinct. Connect the dots of Leo's two brightest stars in the sickle, Algieba and Regulus, and continue that line of sight upward another 23 degrees. You will run right into Hydra's one bright star, Alphard, which means "the solitary one" but was also nicknamed "the backbone of the serpent" by some Arab astronomers. Hydra's heads will look like a little ring of five faint stars at the westernmost (or farthest left) part of the serpent, and its tail will wind over to the right.

Note that although Hydra is huge, it can be a challenge to locate if the night sky has a lot of light pollution. But if you look carefully, Alphard should appear to be a little redder than the other stars in the spring sky. If you are viewing from a small city or out in the country, then you should have no trouble making out Hydra's distinctive and multifaceted heads. The head stars can set long before the heart and tail of Hydra. The ring of stars marking Hydra's head are prominent from February to May while Alphard, the backbone of the serpent, shines from its solitary position from March until June. You can best see the entire Hydra in the evening hours during the months of April, May, and June when it sits very high across the northern sky.

CORVUS, THE CROW

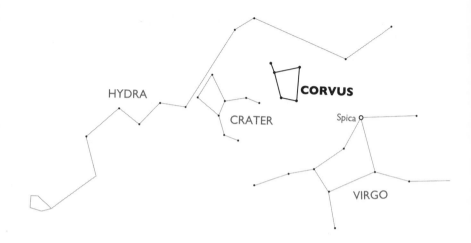

What Is It?
Constellation

Difficulty Level
Moderate

Description
In Greek mythology, one day Apollo worked up a mighty thirst. Apollo was the god of just about everything—from music to truth, and light to healing. But he was best known for guiding the celestial horses and chariot that carried the Sun daily across the sky. Since Apollo barely had any free time to himself, he often employed a crow named Corvus to act as a personal assistant.

When Apollo grew thirsty from carrying the Sun around all day, he sent Corvus with his favorite cup to fetch water from the river.

Several cultures, including the Greeks, looked upon crows as intelligent and crafty creatures. Corvus was no exception. It was such a beautiful sunny day (thanks to the work of Apollo) that Corvus stopped along the way to eat some juicy figs and take a little nap. When he woke up, the day was almost over and the Sun had almost set, and he knew he was going to feel Apollo's wrath for being so late to return. Corvus crafted what he thought was a devious excuse. He filled up the cup with water, picked up a water snake from the river, and then flew back to Apollo. Corvus explained, "Sorry I'm late, Apollo. But this water snake wouldn't let me get any water. So I brought him here to show you I wasn't lying." Apollo, seeing through the horrible falsehood, threw the crow, cup, and snake into the sky to become the constellations Corvus, the Crow; Crater, the Cup (a much fainter nearby constellation); and Hydra, the Many-Headed Snake. Now, Corvus must crawl below the snake for eternity with the cup of water just out of his beak's grasp, making him forever thirsty. This legend was said to explain why earthly crows are cursed with such rough, raspy voices.

How to Find It

Although Corvus is a small constellation it is moderately easy to find in the sky. Look for a trapezoid shape of four semi-bright stars just below the back of Hydra. Corvus, the Crow is closer to the tail end of Hydra and is only a short distance (about 17 degrees) away from Spica, the bright blue star that lies in the constellation Virgo, the Maiden. Corvus will rise in the eastern sky in February and March with Spica and Virgo following soon after. In June the trapezoid shape of Corvus stands so high in the northern sky that it is almost overhead. In July and August it is still very high in the northwestern sky and will get lower night after night until it gets too low in the west to see in September.

CORONA BOREALIS, THE NORTHERN CROWN

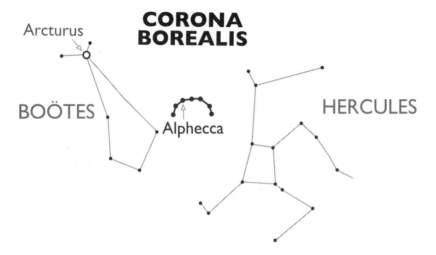

CORONA BOREALIS

Arcturus

BOÖTES

Alphecca

HERCULES

What Is It?
Constellation

Difficulty Level
Moderate

Description
Corona Borealis, aka the Northern Crown, is one of the most identifiable constellations in the sky. Although small and dim, once you find the outline of the seven stars that form a subtle semicircle of sparkling jewels, you will always remember it.

The Greeks likened this star picture to the crown presented to a beautiful maiden named Ariadne who fell in love with Theseus, the prince of Athens. Unfortunately for their love affair, Theseus was chosen to be sacrificed in the great labyrinth of King Minos, whose twisting halls were roamed by the vicious Minotaur—a creature that was half man, half bull. Before Theseus was thrown into the labyrinth, Ariadne gave him a sword to kill the beast and a huge spool of thread. Theseus tied one end of the thread to the entrance and reeled out the thread as he walked through the maze. This way, he would be able to find his way out. Theseus succeeded in slaying the Minotaur and followed the thread back to Ariadne's arms.

The couple lived happily for only a few years, then Theseus grew bored with family life and left Ariadne for her sister, Phaedra. The god of wine, Dionysus, took pity on Ariadne for this act of treachery and granted her the most beautiful crown in the world. Upon her death the crown was placed in the skies for all to see.

This formation of stars is so distinct that many other cultures envisioned an image in this region of the sky. For example, the Australian Aborigines thought this constellation was a boomerang flying through the heavens.

How to Find It

This constellation is not incredibly bright so do not expect it to stand out like Orion's Belt or shine as bright as the stars in the Autumn Triangle. Corona Borealis is a small, subtle grouping of stars but once your eyes alight on them, you will discover a semicircle of stellar gemstones like no other in the northern sky.

Corona Borealis resides between Boötes, the Bear Driver and the mighty winter constellation Hercules. One way to find the crown is to connect a line between Arcturus, Boötes's brightest star, and the center of the keystone shape of four stars that form Hercules's body. Corona Borealis will be halfway between the two features.

Corona Borealis rises after Boötes in the northeast, so you probably won't notice this constellation until May or June. July and August evenings see the crown at its highest point—about halfway up in the northern sky. In September, it will be lower in the northwestern sky and it will be gone from the evening skies around October 1.

CANCER, THE CRAB

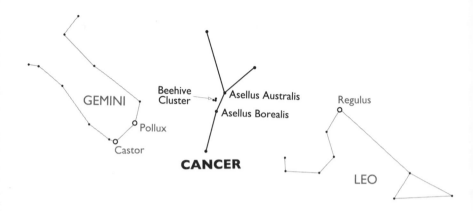

What Is It?

Constellation

Difficulty Level

Difficult

Description

Cancer is a small and faint zodiac constellation visible in the autumn sky. The ancient Greek legend surrounding the constellation is short and sweet. This star pattern is supposed to represent the killer crab sent by the goddess Hera (Zeus's rightfully jealous wife) to harass Hercules, the strongest guy in ancient Greece (and the illegitimate son of Zeus). The plan was to have the crab distract the mighty warrior during his battle with the seven-headed Hydra. However, before the crab could nip at his heels, Hercules made short work of it with one mighty step. Cancer, the Crab was squished underfoot and vanquished to the sky to be honored along with Hercules's other conquests, Leo, the Lion and Hydra, the Many-Headed Snake—an inglorious end for a minor player in the Hercules saga. We will meet Hercules the constellation in the winter sky.

Since Cancer is a zodiac constellation, planets will often wander through these stars. In fact, you may notice the presence of a planet or two much more than the dimmer stars of the crab.

How to Find It

The brightest part of Cancer, the Crab comprises third and fourth magnitude stars that form a Y shape, with the stars Asellus Borealis and Asellus Australis in the center. They definitely will not stand out in your sky and may even be completely invisible if you're viewing from an urban location. The best way to find the area where Cancer resides is to use the more brilliant zodiac constellations on either side of it. Leo, the Lion is to the right of Cancer, and the Gemini Twins are on the left. Identify the upside-down, backward question mark for Leo's head and the dazzling head stars of Gemini, Pollux, and Castor. About 40 degrees of sky separate Regulus and the twin bright stars in Gemini, and Cancer sits about halfway between these two landmarks.

Cancer rises earlier than any of the other autumn constellations, so you can start to see it low in the west in late January and about 30 degrees up in the northeastern sky in February. During those months Pollux and Castor will be hanging 15 degrees to the left of Cancer. The best months to see Cancer in the evening are March and April, when the crab is about halfway up in the northern sky with Leo's brightest star, Regulus, 20 degrees to the right of it. By May, Cancer will be about 30 degrees up in the northwest and will sink lower in the western sky each night until it exits the evening sky altogether in late June.

THE BEEHIVE CLUSTER

What Is It?

Star cluster

Difficulty Level

Difficult

Description

Within the constellation Cancer you can find a gorgeous group of stars. Some people say that the light from these stars resembles a swarm of bees circling a hive, and this grouping was therefore named the Beehive Cluster. This open star cluster consists of 1,000 stars with a common center of mass.

Known from antiquity, the Beehive Cluster inspired many astronomers and writers. Around the year 265 B.C. the Greek author Aratus described it as a "little mist." The Greek astronomer Hipparchus called the cluster a "little cloud" in 130 B.C. And about 200 years later the great astronomer Ptolemy, whose writing guided astronomers for the next 1,400 years, included it as one of the seven nebulas he could see with the naked eye. Some ancient Greeks and Romans also called this group of stars the Praesepe, or "the Manger," with two donkeys—the stars named Asellus Borealis and Asellus Australis—resting among the stars nearby.

How to Find It

The Beehive Cluster shines with an apparent brightness of a third or fourth magnitude star (much dimmer than the stars in Crux, the Southern Cross). So you will need a dark sky to see it clearly. If you can make out the Y shape of Cancer, then look for two stars at the junction of the crab's extremities. These are Asellus Borealis and Asellus Australis, the donkeys. Just to the left of the two donkey stars, if you squint you might, just might, be able to see a small, faint, and fuzzy patch of sky. That is the Beehive Cluster. You may even be able to detect it much better if you don't look directly at it. Astronomers call this averted vision, when you allow your peripheral vision to see faint objects more clearly than if you stare straight at them.

The Winter Sky

As autumn turns to winter you will quickly notice that the Sun is setting earlier, and you have fewer hours of daylight and a lot more darkness. For stargazing that is a great thing because not only do you have more hours to view the night sky but you can start doing it earlier in the day.

And there are so many stars, constellations, and mythical creatures to see in the winter sky that you'll need the extra time to find them all. During these long nights you look for not one but two Centaurs crossing the skies, Centaurus and Sagittarius. Cygnus, the Swan actually looks like a swan with a long neck, beak, and outstretched, starry wings. The tiny constellation Delphinus, the Dolphin leaps out of the cosmic ocean, back arched, with the Milky Way sparkling above. Scorpius, the Scorpion, with its red beating heart star, Antares, climbs high into the night sky as winter progresses.

The stories of the winter constellations are mostly Greek in origin and so were to be imagined by stargazers in the Northern Hemisphere. As mentioned earlier, the view of the stars from the Southern Hemisphere is very different. First off, most of the classical Greek constellations appearing in the winter sky are almost completely upside down when viewed from the Southern Hemisphere. You may not be able to tell when a swan is flying or a scorpion is crawling upside down, but the perspective does affect one of the centaurs (Sagittarius) and a big guy holding a snake (Ophiuchus), who both cross the southern sky standing on their heads. Strangely enough, the constellation Hercules is right-side up in the Southern Hemisphere.

Secondly, since the seasons in the Southern Hemisphere are opposite those in the Northern Hemisphere, these "winter" stars were associated with the summer season in Greece. What I call the "Winter Triangle" in this chapter is an informal star pattern regularly called the "Summer Triangle" in the Northern Hemisphere. Your different latitude on the globe gives you a unique perspective on these stars and allows you to observe additional stars the Greeks could not. Make the most of these long winter nights and come stargazing.

CENTAURUS, THE CENTAUR

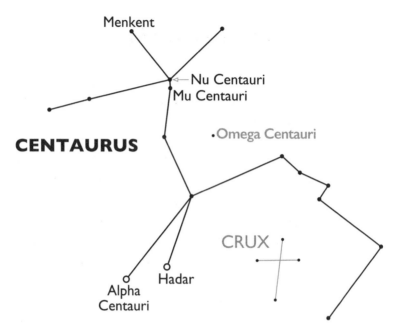

What Is It?

Constellation

Difficulty Level

Easy

Description

Centaurus is a huge constellation spanning an area of the sky 35 degrees wide and 25 degrees high. It contains many gems of the southern sky including Alpha Centauri, our closest star system; several galaxies; and Omega Centauri, the largest globular cluster of stars known in the Milky Way galaxy.

The ancient Greeks must have been fascinated by centaurs, since this is one of two constellations they named for the mythical creatures that were half man, half horse (the other being Sagittarius, which is also visible in the winter sky). Some ancient Greeks considered this centaur to be Chiron, the most wise, learned, and civilized of the centaurs, while other groups attributed these stars to a stereotypical wild, brutish, and mean-spirited centaur (yes, they had stereotypical mythical creatures). In many depictions of Centaurus he is thrusting a spear into a wolf (the constellation Lupus, which is just to the centaur's left). The stars of Lupus are relatively bright and reside between Centaurus's outstretched arm and the curve of Scorpius, the scorpion's tail.

Centaurus is now primarily a constellation visible in the Southern Hemisphere, so how did the ancient Greeks even know about it? Thousands of years ago the Earth's tilt was aimed in a slightly different direction. The Earth has a slow, 26,000-year wobble called precession that changes an observer's perspective on the stars over its 26,000-year cycle. In ancient Greece, the precession made the dynamic southerly stars in Centaurus visible just above their southern horizon.

How to Find It

The easiest way to find this centaur is to first locate the Southern Cross, which stands between its feet. Ten degrees to the east of the Southern Cross you will find two of the brightest stars in the sky, first Hadar (also known as Beta Centauri) and then Alpha Centauri. Alpha and Beta Centauri mark the front feet of the centaur, with the back legs on the other side of the Southern Cross. If you can picture the body of Centaurus using the dimmer stars above the Southern Cross you can also find his semi-bright head star named Menkent. Just connect a line between the two brightest stars in the Southern Cross, from Acrux to Mimosa, and keep going through the centaur's body another 26 degrees and you'll reach Menkent.

Look for Centaurus high in the southern sky (nearly straight overhead from some latitudes) in the winter evenings. By the spring the great beast will appear tipped over about halfway up in the southwest. Warm summer evenings are free from the Centaur's presence since it is below the southern horizon. But by autumn, Centaurus can be seen about halfway up in the southeast and rising higher each night.

ALPHA CENTAURI, THE CLOSEST STAR SYSTEM

What Is It?
Star system

Difficulty Level
Easy

Description
To the naked eye, Alpha Centauri looks to be the third brightest star in the entire sky. Only Sirius in the constellation Canis Major and Canopus in Carina are brighter. But upon closer examination, astronomers discovered some very important things about it. First, Alpha Centauri is not just a solitary star. It is in fact three stars that orbit one another—two yellow suns and one little red one named Proxima Centauri. If you lived on a planet around there, like the one astronomers found orbiting Proxima Centauri in 2016, you'd have three suns in your sky.

What makes this discovery even cooler is that Proxima Centauri is also the closest star to the Sun. It is 4.2 light-years from us, but that is still the equivalent of 40 trillion kilometers. If you wanted to fly to a planet around Proxima Centauri with the fastest spacecraft we ever made, it would still take you more than 70,000 years to get there!

How to Find It
First, face south and locate the Southern Cross. Connect the dots, from right to left, on the two stars marking the shorter line on the cross. Continue that line almost another 10 degrees and you'll come to a very bright star. That is not Alpha Centauri but instead its neighbor Hadar, or Beta Centauri. When you hop another 4 degrees past Hadar you will easily spot an even brighter star and that is Alpha Centauri.

Unlike the majority of the constellation Centaurus, Alpha Centauri and Hadar are visible nearly every night of the year. Although they are highest in the southern sky during the winter evenings, you can still see them in the southwest during the spring and the southeast during the autumn. Only during the summer months can Alpha Centauri dip low and sometimes below the southern horizon and not be visible.

OMEGA CENTAURI, SEE 10 MILLION STARS

What Is It?
Globular cluster

Difficulty Level
Moderate

Description
The Milky Way galaxy houses about 150 known globular clusters. Think of them as huge conglomerations of stars that are jam-packed together into a globe shape. Omega Centauri, which is 15,000 light-years away from Earth, is the largest globular cluster in the Milky Way. It is 170 light-years wide, has over 10 million stars, and weighs in at more than 4 million Suns.

Omega Centauri is not visible from most of the Northern Hemisphere but looks amazing from the Tropics and any location in the Southern Hemisphere. It is rated as a third magnitude object, meaning you can see it from rural and suburban locations. At first it will look like a tiny cloud, but as you peer deeper you might make out the structure as a tiny globe of sparkling lights. Omega Centauri looks just awesome through a pair of binoculars, where the mass of stars takes on an almost three-dimensional quality. And in a telescope, this globular cluster can fill the entire eyepiece with countless stars.

How to Find It
To find Omega Centauri, you need to find Centaurus's head star, Menkent. Just connect a line between the two brightest stars in the Southern Cross, from Acrux to Mimosa, and keep going through the centaur's body another 26 degrees and you'll reach Menkent. Then trace a line from Centaurus's head star Menkent to two dimmer stars named Mu and Nu Centauri that are about 6 degrees away. Continue that line for another 7 degrees and you will find Omega Centauri.

Omega Centauri stands highest above the southern horizon every May, June, and July evening. During these winter months, Omega Centauri can be more than 60 degrees up in the south, so it may help to recline in a chair to see it. Every August and September, Omega Centauri circles lower and to the right and is about halfway up in the southwestern sky. From October through February, this cluster is difficult to find right after sunset because it is below the southern horizon for most viewers. But by March and April, it returns to the evening sky, sitting about halfway up in the southeast.

THE WINTER TRIANGLE

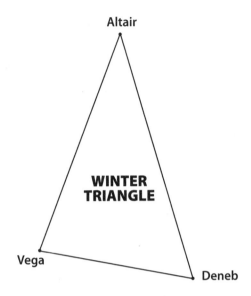

Altair

**WINTER
TRIANGLE**

Vega

Deneb

What Is It?

Asterism

Difficulty Level

Easy

Description

In July, August, and September when you face north you'll find a triangle of three celestial dazzlers—stars named Vega, Deneb, and Altair. When you connect these stars with lines, they form a large triangle measuring about 30 degrees long and 20 degrees high. Remember the easy way to measure degrees in the sky? Your fist at arm's length is about 10 degrees. So this shape in the sky will be two fists high and three fists long.

Southern Hemisphere astronomers call this formation the Winter Triangle, which, like the False Cross and the Autumn Triangle, is an asterism (a recognizable shape of stars) rather than an official constellation. In fact, each star in the triangle is part of its own constellation. Vega is the brightest star in the constellation Lyra, the Harp. Deneb is the tail star of

Cygnus, the Swan. And Altair is the eagle eye of the constellation Aquila, the Eagle. The Winter Triangle is super easy to identify, and you'll see all three stars at some point in the night in the wintertime, even if you live in the heart of a city or under any light-polluted sky.

How to Find It

This asterism is called the Winter Triangle because its three stars, Vega, Altair, and Deneb, become really noticeable in the heart of the winter season. On July evenings you can find it just above the northeastern horizon. Vega and Altair rise almost simultaneously, with Vega, the brighter of the two stars, on the left and Altair to the right. Deneb rises last and is the farthest north of the three stars. That means Deneb, in its nightly arc across the northern sky, will stay the lowest above the horizon and be visible the shortest amount of time of the three stars in the Winter Triangle. In September the Winter Triangle is visible due north, with Vega and Deneb at the base of the triangle and Altair sticking straight up in the sky. You can still see it in October as it starts to tip over in the northwestern sky after dark. Vega and Deneb will set on October evenings while Altair will remain in the sky through November and will set more toward the western horizon.

LYRA, THE HARP

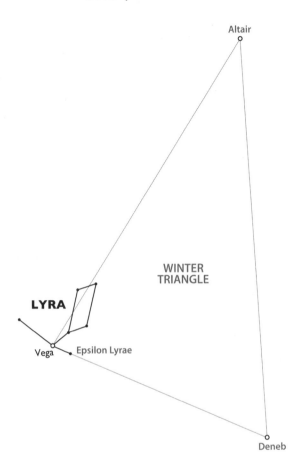

What Is It?
Constellation

Difficulty Level
Moderate

Description
According to some ancient Greek legends, Lyra was the harp of Orpheus, the best musician who ever walked the Earth. His music was so sweet and pure that the trees bent over to listen, the rivers ceased flowing, wild beasts became tame, and even mountains listened with pleasure when

Orpheus played his magical music. One day Orpheus met a beautiful nymph named Eurydice. She, too, was mesmerized by Orpheus's music, and they fell in love and got married. Unfortunately, soon after the marriage Eurydice was bitten on the heel by a serpent and died. Orpheus was deeply saddened by her death and vowed to never play music again.

The gods, missing his sweet music as much as any mortal, came to Orpheus and allowed him safe passage to Hades to retrieve his lost love and bring her back to the land of the living. Through song, Orpheus convinced the god of the underworld to release Eurydice on the condition that he could not look at her until they emerged into the land of the living. But as they left Hades, Orpheus couldn't help himself and he looked back for Eurydice and saw her being dragged back to the underworld. Orpheus's harp, the constellation Lyra, now remains in the sky as a reminder of true love, love lost, and why death is so hard to cheat.

Although small, this constellation holds many bright stars, including the fifth brightest star in the sky, Vega. In the heart of winter you will find Vega sparkling like a blue-white jewel at the left corner of the Winter Triangle. Four fainter stars form a neat little parallelogram that seems to hang off Vega by an invisible string. The parallelogram is the body of the harp and so "strings" are definitely appropriate imagery for this star pattern.

How to Find It

To locate the harp's position simply find Lyra's brightest star, Vega. From a darker sky the outline of the harp is more apparent. You may also see a sixth star with the naked eye close to Vega. That is called Epsilon Lyrae and is a famous quadruple star system. Astronomers also call Epsilon Lyrae "the Double," since you can see two tightly knit pairs of stars through a telescope. When you can make out the entire harp in the sky, the view of this tiny constellation is as sweet as Orpheus's music.

In mid-winter Lyra rises in the northeastern sky after dark. The months of August and September see the harp constellation reach its highest points in the northern sky but it is still only about 20–30 degrees above the horizon. By October Vega and the whole constellation are just above the northwestern sky during the evening.

VEGA, THE STAR OF WINTER

What Is It?
Star

Difficulty Level
Easy

Description
Worshipped in ancient Egypt as early as 6000 B.C., and with its prime location in the northern sky, Vega is definitely the star of the winter nights.

Vega is a blue-white giant star that burns at well over 9,300°C on its surface—much hotter than our 5,500°C yellow Sun. Vega is one of our closer stars, only about 25 light-years from Earth. Our solar system is constantly moving in the direction of Vega at a rate of almost 20 kilometers per second. Even though the Sun, Earth, and all the planets are traveling through space at tremendous velocity, it will still take over 500 million years for our two stars to pass close to one another.

Vega is so bright that for some Polynesian groups, its appearance marked the beginning of their new year. For the Boorong people of Australia, the star Vega marked the position of a bird constellation called the Neilloan. The name Vega comes from the Arabic for "the falling eagle," and other stargazers in the Middle East considered this star to be part of a vulture lurking in the night sky.

How to Find It
Vega is the brightest star in not only the constellation Lyra but also the asterism of the Winter Triangle. It's a safe bet that when you face north on any given winter evening, Vega—with the exception of possibly a planet—will be the brightest twinkling celestial body you see. You can first spy Vega late in the month of June in the northeastern sky after dark. It forms the left corner of the Winter Triangle in July, August, and September and sits about a third of the way up in the northern sky. Once October rolls around Vega will get lower in the northwest and be gone from the evening skies soon after.

CYGNUS, THE SWAN

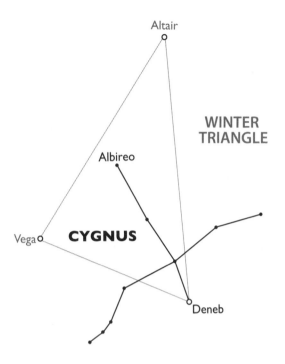

What Is It?
Constellation

Difficulty Level
Easy

Description
One Greek legend describes Cygnus as the chief god himself, Zeus. One day Zeus fell in love with the queen of Sparta, a mortal woman named Leda. To woo the queen, Zeus disguised himself as a swan. Each day this strange yet compelling swan would appear at Leda's window. Eventually she thought, "My, there is something very attractive about that swan." Maybe it was the feathers or the way he threw thunderbolts, but as strange as it sounds, Leda fell in love with the swan. From this divine yet unholy union Leda became pregnant and, yes, laid an egg. When the egg cracked, out popped the Gemini twins, Pollux and Castor.

Cygnus, the Swan is one of the easier constellations to picture in the sky. It does not take a lot of imagination to see a bird flying through the heavens in these stars. Start from the tail star Deneb and move inside the Winter Triangle to find fainter stars in a line that form the body and two windswept wings. A star named Albireo represents the swan's head at the end of a long neck.

Instead of a swan many Northern Hemisphere stargazers have given Cygnus a nickname: the Northern Cross. The line between the stars Deneb and Albireo is the long part of the cross, while the line of three stars making the swan's body and wings form the shorter part of the cross. For Southern Hemisphere observers, the Northern Cross is in the northern sky but is upside down and may look more like a swan with its tail just above the horizon than a letter "t."

The swan is located in a rich part of the Milky Way. In some early cultures the Milky Way was considered to be the river in the sky and the swan, a water bird, naturally flew along it. The Milky Way appears to split into two channels in Cygnus with a dark, apparently starless area in the middle. That darker rift in the Milky Way is called the Coalsack.

How to Find It

To find Cygnus, look for its brightest star, Deneb, securing the lower-right corner of the Winter Triangle and indicating the position of Cygnus's cross-like shape. By late July and early August Deneb and Cygnus anchor the lower right side of the Winter Triangle when it rises in the northeastern sky.

Cygnus doesn't get very high in the northern sky, so you only have a short window to see it. The best months to spy it are September and October when it sits just above the northern horizon. You may need to find a viewing location free from buildings and trees to see it flying just above the ground.

DENEB, THE TAIL OF THE SWAN

What Is It?
Star

Difficulty Level
Easy

Description
The name Deneb comes from the Arabic word for "tail." This star shows you the tail end of Cygnus, the Swan.

Deneb shines with a blue-white light and scorches space with a surface temperature of about 8,200°C. Not only is it the farthest star in the Winter Triangle; Deneb is perhaps the farthest star you can see with the naked eye. Although sky surveys vary in their measurement of Deneb's distance from Earth, some astronomers think it could be more than 3,000 light-years away. That's the equivalent of about 27,000,000,000,000,000 kilometers! The fact that we can still see it as a bright star must mean that Deneb is humongous. By some estimates Deneb could be about 19 times more massive, over 200 times wider, and 200,000 times brighter than our Sun. If Deneb were hundreds of times closer to Earth—like at the equivalent distance in which Sirius resides—it would completely light up the nighttime sky.

Because of its great distance from us, Deneb appears to be the faintest star in the Winter Triangle and is also the farthest north of the three great brilliant stars of this asterism.

How to Find It
While the brighter star Vega rises in the northeast in July, dimmer Deneb emerges from the north-northeastern horizon a few hours later and marks the lower-right corner of the Winter Triangle. It will be about 24 degrees to the right of Vega. Deneb is so far north that you have to look for it at the right time from the right place. It barely creeps above the northern sky all winter and into the spring. The best months to observe Deneb are September and October when it shines just above the northern horizon.

AQUILA, THE EAGLE

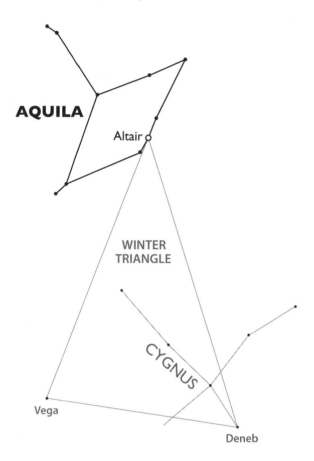

What Is It?
Constellation

Difficulty Level
Moderate

Description
Many Greek and Roman myths indicated that Aquila, the Eagle carried the thunderbolts that the gods Zeus or Jupiter frequently hurled at troublesome humans. In fact, since eagles seemed to be the fastest creatures, the Greeks associated them with lightning strikes.

The Greeks created several myths around this bird of prey: some stories gave the eagle very mundane tasks to do while other tales were downright terrifying. In one legend Zeus was sitting on Mount Olympus and developed a mighty thirst. Instead of getting a drink himself, he sent his eagle to fetch a mortal from below to do the job. The eagle raced down to the city of Troy and snatched up a young prince named Ganymede and brought him back to Mount Olympus. Ganymede, far removed from being a prince, then became a servant to Zeus as his cupbearer.

In another tale, the eagle became an instrument of torture. The god Prometheus developed a soft spot for humans and how they suffered on Earth. He decided to introduce fire to humanity and help them get through bad weather. Zeus, fearing the humans would now become too powerful, punished Prometheus for this act by chaining him to a rock. When the Sun rose each morning, Zeus sent his eagle to sit on Prometheus's lap and eat out his liver. At the end of the day, the eagle would fly home and Prometheus's liver would magically grow back. As the Sun rose, the eagle returned, and this punishment went on day after day for eternity.

Beyond the dazzling star Altair, Aquila has very few bright stars, and its outline is tough to discern. It may be drawn as a diamond or a chevron shape, depending on the artist.

How to Find It

The easiest way to find Aquila is to locate Altair in its far-flung position in the Winter Triangle. While Vega and Deneb are only 24 degrees apart, Altair is the farthest star from the other two (34 degrees from Vega and 38 degrees from Deneb). Aquila is also the southernmost member of the Winter Triangle. This means it will travel much higher in the sky and be visible much longer than the other two stars. At the beginning of winter, look for the eagle rising in the east-northeast just after dark. Don't confuse Altair with brighter Vega, which will be rising at the same time 34 degrees to the left of Altair. In July and August

Altair will sit about halfway up in the northeastern sky. In September it flies highest in the heavens atop the Winter Triangle and about 60 degrees above the northern horizon. In October it is still relatively high in the northwestern sky and will remain long after Vega and Deneb depart, so by November you may only see Altair alone while the other corners of the triangle have set.

ALTAIR, THE EAGLE EYE

What Is It?
Star

Difficulty Level
Easy

Description
Altair is only 17 light-years away from Earth and is so close (relatively speaking) that we can actually study it in better detail than most stars in the galaxy. Using the best telescopes in the world, astronomers are able to see that Altair completes one turn every nine hours (as opposed to the Sun's every 25 to 26 days). This rapid rotation has flattened it at the poles and sent more mass to its equator, which makes it look more like a blue-white egg than a sphere.

An old Korean legend links Altair to the star Vega in the constellation Lyra. In this story two star-crossed lovers, a cowherd and a weaver, were banished to the heavens. It looked as if they would finally be together until a flock of magpies flew in between them and they glided away from each other. When they stuck to the heavens, the weaver (Vega) was on one side of the great river in the sky (the Milky Way) and the cowherd (Altair) fell on the opposite shore.

The couple could meet only once a year on the seventh day of the seventh month. Only then could they cross the river with the help of friendly magpies. Magpies in Korea flocked like crazy in July and were said to fly up to the stars to construct a bridge across the river for the couple to cross. Legends say that when the meeting occurs on July 7, Altair and Vega shine in five colors to symbolize their happiness. If it rains on July 7, so the legend goes, it is a sad night, for the couple fails to meet at all.

How to Find It
Altair is the eagle eye of the constellation Aquila, the Eagle and is the second brightest star in the Winter Triangle. Once you identify the Winter Triangle, Altair is the star farthest from the other two and appears highest in the sky.

DELPHINUS, THE DOLPHIN

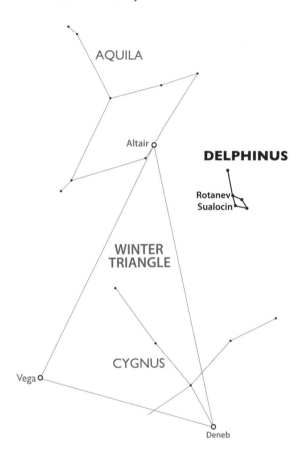

What Is It?

Constellation

Difficulty Level

Moderate

Description

Delphinus, the Dolphin is one of the smallest constellations in the winter sky, and it lies just outside the Winter Triangle. Once you find it, you may even be persuaded to agree it looks a little like a dolphin arching its back and jumping above the cosmic waves. According to Greek myths, Delphinus was said to be the messenger of Poseidon, the god of the sea.

Delphinus won great acclaim for saving the life of Arion (Poseidon's son) when his ship was attacked at sea. Ancient mariners often attributed to the dolphin great wisdom and also a love for humans. Dolphins were thought to be sailors' best friends, and legends (both real and fictional) swirled through the Mediterranean about how they had rescued humans and gotten shipwrecked fools out of trouble. Delphinus also helped Poseidon get a date with the beautiful Nereid named Amphitrite by delivering her a message when she was in the Atlas Mountains. I'm not sure how this "dolphin-gram" reached her, since he couldn't exactly swim it up the mountain, but that would be an interesting sight!

The great river in the sky, the Milky Way, is not far from Delphinus. The cloudy band of stars flows just above the dolphin's head and spans the area between the constellations Cygnus and Aquila.

How to Find It

Delphinus consists of third and fourth magnitude stars, so it will be a challenge to find this little constellation if you live in a city. Even if you live out in the country, this star pattern won't jump out at you. You will see it out of the corner of your eye as a little faint glow, but upon closer examination you should be able to make out the distinctive arch of stars that covers only about 5 degrees of the summer sky.

To find Delphinus, look along the longest side of the Winter Triangle (between the stars Altair and Deneb). Just outside the triangle and closer to Altair search for a small diamond shape of four stars with one or more stars just above it. The stars in the diamond mark the body, and the extra star above is the tail.

It's easier to spot the Dolphin when it is higher in the sky. That doesn't happen until evenings in August when it rises higher above the northeastern horizon after dark. In September and October you can find Delphinus high in the northern sky almost two-thirds of the way up. By November it hangs out in the northwest and is pretty much gone from the evening sky from December until July.

SUALOCIN AND ROTANEV

What Is It?
Stars

Difficulty Level
Moderate

Description
You have read about stars named by the Greeks, Romans, and Arabs, but the two brightest stars in Delphinus—Sualocin and Rotanev—are a little more modern and are actually named after one particular person, a man named Niccolò Cacciatore.

Cacciatore was an assistant to the Sicilian astronomer Giuseppe Piazzi, the discoverer of the first and largest asteroid, Ceres. In 1814 Cacciatore helped Piazzi prepare a new star catalog. When other astronomers read the work, they found that two stars in Delphinus had acquired new and unique names. Only decades later did anyone figure out that the names, Sualocin and Rotanev, were the reverse spelling of Nicolaus Venator, the Latinized version of the name Niccolò Cacciatore. Niccolò snuck his own name into the stars of Delphinus! And the names have stuck ever since and show a most creative way to become immortalized in the stars.

Sualocin and Rotanev make up the arching back of the dolphin constellation. Sualocin, on the top of the diamond shape, is blue-white in color and about 241 light-years from Earth. Rotanev, on the right side of the dolphin, is a white star about 101 light-years away.

How to Find It
To find Delphinus's uniquely named stars, scan along the longest side of the Winter Triangle (between the stars Altair and Deneb). Just outside the boundaries of the triangle and closer to Altair you may find a small diamond shape that forms the body of the Dolphin. Sualocin and Rotanev are the brightest two stars in the diamond and should shine with a similar white glow. Sualocin marks the top of the dolphin's head, with Rotanev just 1.5 degrees above it.

HERCULES, THE KNEELER

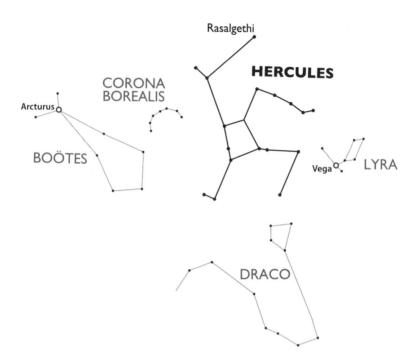

What Is It?
Constellation

Difficulty Level
Difficult

Description
In Greek mythology Hercules (or Heracles in Greek) was the son of Zeus and a mortal woman, Alcmene. His name was a slap in the face to Zeus's wife, Hera, since the name Heracles means "glory of Hera." Through a bit of sorcery, Hera drove Hercules stark raving mad—mad enough to kill his wife and kids. As punishment for his crimes (driven by madness or not) Hercules was forced to do twelve labors for King Eurystheus. At the completion of these labors, including the slaying of the Nemean Lion (the constellation Leo) and the many-headed serpent (the constellation Hydra), Hercules would achieve immortality.

All the stars that make up the constellation of this mythical demigod are dim. Only Hercules's head star, Rasalgethi, stands out, and even it barely shines brighter than third magnitude (only one-fourth as bright as the stars in Crux, the Southern Cross). Rasalgethi is an old Arabic word meaning "head of the kneeler." It has the uniquely red hue of a supergiant star. Rasalgethi's brightness also varies wildly—a common trait for red giants. When you find Hercules's head in the stars, aim a small telescope at Rasalgethi and you will discover that it is really two stars in one. Some ancient Greeks pictured Hercules in the stars kneeling with an upraised club in one hand and Hera's snakes in the other.

How to Find It

The best way to find the bulk of Hercules is to look for the keystone, which is a four-sided figure that makes up the torso of his body. Use the curving star pattern of Corona Borealis to point you to the keystone's location by following the lip of its semicircle of stars from left to right. At the end of the right side, exit the curve and keep going about 14 degrees farther. That will take you to the center of the keystone.

You can also identify Hercules's place in the sky by looking at the more noticeable stars and constellations surrounding him. Corona Borealis will be to his left, while the bright star Vega in the constellation Lyra, the Harp will be about 20 degrees to his right.

Once you note the position of the keystone, you may be able to imagine Hercules's arms, kneeling legs, and head. His outline is totally spread out with legs and arms flailing in all directions. He rises after dark in the northeast in late June and early July. Hercules may be a dynamic character in mythology but he does not stand very high in the sky from the Southern Hemisphere. He is highest on July and August evenings but only reaches about 30 degrees above the northern horizon. As September rolls around, Hercules is barely above the northwestern horizon.

OPHIUCHUS, THE SERPENT BEARER

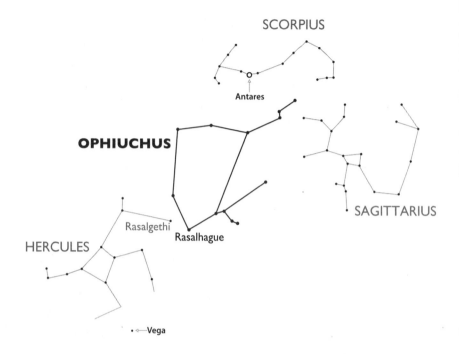

What Is It?
Constellation

Difficulty Level
Moderate

Description

Many legends say that this constellation shows a man holding a huge serpent. The Greeks believed this guy was named Asklepios instead of Ophiuchus. He was the god and inventor of medicine and was so gifted that he could even restore the dead to life.

Asklepios was the son of the god Apollo and a mortal princess named Coronis. He was cared for and mentored in the field of medicine by the centaur Chiron (some say Chiron is embodied in the nearby constellation Sagittarius). Once when Asklepios was making a house call to a sick patient, a serpent slithered into the room and coiled around a staff. Asklepios, slightly scared of snakes, quickly killed it. A few minutes later a second serpent crawled under the door and entered the room carrying an odd herb in its fanged mouth. The second serpent went over to the dead one and applied the herb, restoring the dead serpent to life. From that moment on Asklepios always carried a staff with a serpent wrapped around it. It became the symbol for the medical arts still in use today.

Asklepios brought many people back from the dead, including Orion. As described in the Orion myth, the gods sent a scorpion to humble Orion—by killing him. In one alternate ending of the myth Asklepios was called to the scene of the crime to work his doctorly deeds. Not only did Asklepios raise Orion to life but he even dispatched the scorpion by squishing it under his sandaled foot.

The constellation Ophiuchus looks like a long stretched-out pentagon of stars, but none of them are terribly bright. The star named Rasalhague (meaning "head of the serpent charmer") is the brightest star in Ophiuchus and marks the bottom of the pentagon (he's standing on his head in the Southern Hemisphere). Rasalhague is a white star lying about 49 light-years from Earth. Just to the left of Rasalhague is a star in the constellation Hercules called Rasalgethi, or "head of the kneeler." So, very close together, we have two men bumping heads in the heavens! Hercules gets to kneel upright, while Ophiuchus looks like he's holding a serpent while standing on his head.

How to Find It

To find the elongated pentagon shape of Ophiuchus, search about 25 degrees below the constellation Scorpius and about 25 degrees above Hercules. You can also locate Ophiuchus by finding its brightest star, Rasalhague, which marks the lowermost corner of the pentagon shape. When Ophiuchus stands upright halfway up in the northern sky during the winter months, Rasalhague is bracketed by two very bright stars. Look for Antares in Scorpius, the Scorpion, higher in the north, and Vega from the Winter Triangle below it. If you draw a line from Antares down to Vega (a jump of about 70 degrees), you'll find Rasalhague on that line about 40 degrees below Antares.

The best months to look for Ophiuchus are July and August when he stands about two-thirds of the way up in the northern sky. However, you can find him halfway up in the east-northeast during June and later halfway up in the west-northwest in September.

SCORPIUS, THE SCORPION

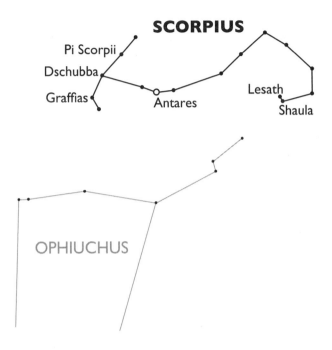

What Is It?
Constellation

Difficulty Level
Easy

Description
Scorpius is a very distinct constellation that crawls up to nearly the zenith on most winter nights. The brightest stars in Scorpius form the shape of a fishhook or the letter "J." This is the body and tail of the scorpion, and out of all the constellations in the sky this group of stars may look the most like its namesake. But as usual, the view from the Southern Hemisphere is upside down from that of ancient Greece so you may have to adjust your imagination.

The fishhook and brightest star, Antares, are definitely Scorpius's most notable features, since the head and claws of the scorpion can be very obscure. The end of the scorpion's tail has two stars, Shaula and Lesath, which are uniquely close together and create the scorpion's stinger.

Mimicking Orion's Belt, Scorpius has three stars in a similar line at its head. The stars are not quite as bright and not quite as aligned as Orion's Belt, but they are definitely noticeable. The star at the bottom of the line is named Graffias, the star in the middle is Dschubba, and the one at the top is Pi Scorpii. The scorpion's claws that reach out to the left are very difficult to imagine unless you're viewing a dark sky. Although astrology refers to this zodiac sign as Scorpio, you should use its more astronomical moniker: Scorpius. This is the killer of Orion, the dreaded scorpion that stung the hunter on his heel and sent him to the afterlife among the stars. Legend has it that after Orion died the gods offered him a place in the sky to live on as a constellation. Orion made one simple request: he never wanted to see that scorpion again. The gods honored this wish and placed him on the opposite side of the firmament from the scorpion. The two constellations are never in the sky at the same time. Orion rules the summer sky, and when he rises in the east Scorpius has already set in the west. During the winter months Scorpius emerges into the sky only after Orion has left the scene and has drifted below the western horizon.

How to Find It

The stars in Scorpius make a very distinct fishhook-shaped pattern in the nighttime sky. It actually first becomes visible in the sky during May when it rises in the east-southeast. Its head of three stars in a row will come up first, followed by its bright red heart star, Antares, and then the curved tail and stinger will trail off to the right. By June, Scorpius will stand about halfway up in the eastern sky, and then you may notice its spot above the stars of Ophiuchus.

In July and August the scorpion constellation will appear so high in the sky that you may want to lie down to watch it. In fact from some mid-southern latitudes, Scorpius can shine from the zenith (straight overhead). In September it is still very high in the western sky but each night afterward you'll find Scorpius diving headfirst lower and lower to the southwest. By the end of October the three head stars will disappear below the west-southwestern horizon, with Antares and the stinger stars following some weeks later.

ANTARES, THE HEART OF THE SCORPION

What Is It?

Star

Difficulty Level

Easy

Description

Antares is one of the reddest stars visible to the naked eye. It also just happens to mark the heart of the constellation Scorpius, and when low in the sky can twinkle, creating an impression of a wildly beating heart in the middle of the scorpion's body.

Antares is a red supergiant star and one of the largest known stars in the galaxy. If Antares were our sun, it would engulf the entire orbit of Mars. Earth would be well inside it. But luckily Antares lies about 600 light-years away.

The name Antares comes from the ancient Greek word meaning "rival of Mars," because of its similar color to the Red Planet. The Chinese called it the Fire Star for the same reason. Occasionally the planet Mars passes near Antares and you can see them side by side. Although they may seem to be the same color, Mars, as a planet, does not twinkle nearly as much as Antares.

As another link to the constellation Orion, Antares has many similarities to Betelgeuse, the star marking the hunter's armpit. Astronomers are watching and waiting for Antares to suddenly flare up into a massive supernova. Which star will blow up first, Antares or Betelgeuse? The answer to that is anyone's guess. But when Antares goes supernova, the light we see from this star may even outshine a Full Moon.

How to Find It

To find Antares, look for the brightest star in Scorpius marking the creature's red, beating heart. Antares is visible in the evening sky between May and October. And during the months of July and August, save your neck and lie back in the grass to find it because it will be nearly straight over your head.

BUTTERFLY AND PTOLEMY'S CLUSTERS

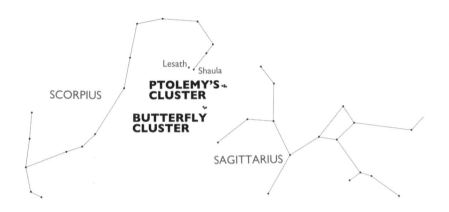

What Is It?
Star clusters

Difficulty Level
Difficult

Description
When you look around the tail stars of the constellation Scorpius under a dark sky, don't be surprised if you detect two faint, blurry patches amid the blackened sky. These are two open clusters, groups of dozens to hundreds of stars in one spot. A famous example of an open cluster is the Seven Sisters cluster, which is visible in the winter sky.

The Butterfly Cluster and Ptolemy's Cluster have been known since antiquity, and they are just barely visible to the naked eye. They are also named M6 and M7, respectively. (They were the sixth and seventh deep space objects catalogued by French astronomer Charles Messier in the eighteenth century, and he labeled them using the initial of his last name.)

Ptolemy's Cluster (M7) is the brighter and closer of the two clusters in Scorpius. The combined light of the 80 stars in this grouping is nearly equivalent to a third magnitude star, and they all lie almost 1,000 light-years from Earth. The cluster is named after the ancient

Greek astronomer Ptolemy, who wrote some of the most important and influential astronomy texts in history. In A.D. 130 Ptolemy described this group of stars as the "nebula following the sting of Scorpius," since it seems so near to the scorpion's stinger stars, Shaula and Lesath. He called it a nebula because he could not clearly see individual stars in the cluster (he lived almost 1,500 years before the invention of the telescope, so his view was limited). To the naked eye it looked more like a small cloudy patch of sky than a star cluster like the Pleiades.

Three degrees to the north of Ptolemy's Cluster, you may find the smaller and fainter Butterfly Cluster (M6). The stars in this group are estimated to be around 1,600 light-years from Earth and are still visible, although faintly, to a well-trained, naked-eye observer. The entire cluster shines at around fourth magnitude, so you will need a dark sky to see it distinctly.

How to Find It

M6 and M7 are best seen during winter evenings and hang out below Scorpius's tail and stinger. First find the constellation Scorpius and its distinctive fishhook shape of stars. Follow the body of the scorpion down and to the right until you reach the two stinger stars that seem incredibly close together, Shaula and Lesath. Ptolemy's Cluster (M7) is about 4 degrees to the right of these stars. The Butterfly Cluster (M6) may be a tougher challenge to spot, but at best you may be able to make it out as a little cloud just below and to the left of M7.

The two clusters look much more dramatic through a pair of binoculars or a small telescope, and by using them you can resolve individual stars quite clearly. Hunt around to the left of the scorpion's stinger and you should see them like a small swarm of fireflies in the winter night. Does the shape of M6's stars vaguely look like a butterfly? You be the judge.

LIBRA, THE SCALES

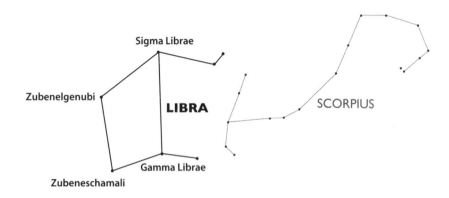

What Is It?
Constellation

Difficulty Level
Moderate

Description
Libra was the last zodiac constellation added to the twelve and is the only one that is not a living creature. This diamond-shaped star pattern is supposed to represent the scales of justice that maintains the balance of law and order. The constellation came into vogue in ancient Rome when the ruling class had supreme respect for maintaining the rule of law.

The symbol for justice is a woman holding a set of scales. This may have come from a combination of Libra and the nearby constellation Virgo. In Greek mythology Astraea was the goddess of justice and sometimes stood in for Virgo among the stars. But according to classical illustrations of Virgo, the scales of justice can be found at her feet and not in her hand.

Scorpius is also near Libra, and the claws of the scorpion can be imagined pinching at the scales. Thousands of years ago the stars of Libra were often incorporated into the outline of the scorpion constellation, and Libra's brightest stars, Zubeneschamali and Zubenelgenubi, reflect this ancient connection. Zubeneschamali is the slightly brighter of the pair and is located lower in the sky. Its name means "the northern claw." Zubenelgenubi, standing above its fellow Libran star, means "the southern claw." In fact, the entire constellation of Libra used to mark the claws of a much larger scorpion. Only later did astronomers break Libra free from the scorpion's grasp and create a balance of power in the heavens.

How to Find It

Look for this small constellation 10–15 degrees to the left of Scorpius. Libra's four main stars form a diamond shape, with Zubeneschamali at the bottom and Zubenelgenubi standing on the left corner of the diamond. Zubeneschamali and Zubenelgenubi are significantly brighter than the other two stars in the diamond shape, Gamma Librae and Sigma Librae, so you may not see the entire shape of Libra right away. But if you fix your gaze to the left of the scorpion you may still picture the stars of Libra as the scorpion's claws.

Libra will rise before the scorpion and start to be visible low in the eastern sky by late April. By May it is about halfway up in the northeast and will soon climb to the roof of the heavens. Like Scorpius, Libra can also appear almost straight overhead but gets there in June and July, about one month earlier than the Scorpion. And in August and September Libra weighs the stars halfway up in the western sky and sets with the Sun in November. The constellation is then not visible in the evening skies between November and March.

SAGITTARIUS, THE ARCHER

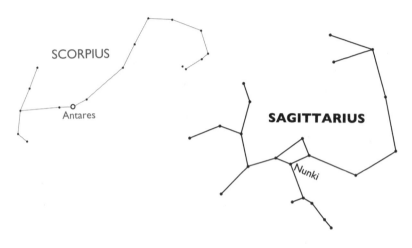

What Is It?

Constellation

Difficulty Level

Moderate

Description

The constellation Sagittarius represents Chiron, the Centaur, who is half man, half horse. Unlike most mythological centaurs that were wild, rude, and crude, Chiron was gentle, kind, and intellectual. His thirst for knowledge led him to become a great teacher of just about everyone in ancient Greece, including Hercules.

The shape of Sagittarius's most notable stars resemble an upside-down teapot more than a creature that is half man, half horse. Like the fishhook pattern of stars that comprise Scorpius, the Scorpion, the teapot is another identifiable asterism—stars that are part of a greater constellation.

The star pattern of Sagittarius can be broken up into two smaller sections: the Bow and Arrow and the Milk Dipper. Look to the spout of the teapot pointing to the left to find Sagittarius's bow and arrow. Three stars curve to form the bow and one sticks out to form the arrow. Look at what he's aiming that arrow at! It is pointed directly at the scorpion's heart, the red star Antares.

The four stars that comprise the handle of the teapot on the left side of the star pattern also look a little like a dipping spoon. Some astronomers granted this formation the additional name "the Milk Dipper" because it lies in a thick patch of the Milky Way. When you are looking at Sagittarius, you are peering toward the center of our galaxy. Some imaginative stargazers picture the Milky Way as steam coming out of the spout of the teapot.

Sagittarius does not have any super bright stars, but try to locate Nunki, the brightest star in the Milk Dipper and the second brightest star in the constellation. It was named by the ancient Sumerians about 5,000 years ago, but today we have no idea what Nunki means. Maybe after observing this blue-white mystery star you can invent your own translation.

How to Find It

If you can find Scorpius, you can find Sagittarius. Look for him prancing high across the northern sky by mid-winter, 15 degrees to the right of Scorpius's stinger.

Sagittarius trails behind Scorpius in its nightly motion across the sky. The centaur rises a little later than the scorpion, and you can get a glimpse of the teapot shape of stars as early as June in the eastern sky. Sagittarius is most easily spotted in August and September when it reaches its highest point above the northern horizon and can even reach the zenith. By October the point of his arrow starts turning toward the western horizon and each night thereafter this constellation will appear lower in the western sky. But Sagittarius can linger on and still be seen through the month of November just above the west-southern horizon.

The Spring Sky

As the month of September arrives you will find the stars and constellations of winter shifted ever westward in the sky. You will still find the three stars of the Winter Triangle, with Vega and Deneb near the northwestern horizon and Altair standing on high and pointing toward the zenith. Scorpius, the Scorpion is still visible high in the southwest with Sagittarius, the Archer above it still aiming its notched arrow at the scorpion's heart. But when you look east after dark a new group of constellations has arrived on the scene and has crept above the distant horizon. The stars of spring are not as bright as those in winter. Apart from the dazzling first magnitude star named Fomalhaut in the constellation Piscis Australis, the spring constellations are primarily constructed with second and third magnitude stars whose muted brilliance can be a challenge to see from a city.

If your imagination is strong you can observe among these fainter stars of spring a mythological epic complete with heroic monster battles, a maiden unchained, and a flying horse. Each night you can look for our hero, Perseus, who flashes the severed head of Medusa to defeat the sea monster Cetus and rescue the fair princess Andromeda. These three constellations, Perseus, Cetus, and Andromeda ride across the spring sky together as if borne on the wings of a flying horse. And right on cue, the extensive constellation Pegasus, the Flying Horse spreads its wings across this season's stars and serves as an easily identifiable landmark. From Pegasus you will then soar over to the other star patterns like Piscis Australis, the Southern Fish; Grus, the Crane; and Aries, the Ram and enjoy a warm evening under the spring stars.

PEGASUS, THE FLYING HORSE

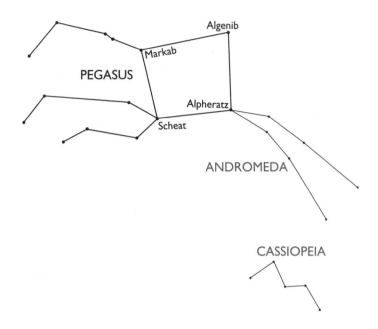

What Is It?
Constellation

Difficulty Level
Easy

Description
One of the most iconic creatures from Greek mythology is the beautiful, white, winged stallion, Pegasus, the Flying Horse. Where did Pegasus come from? This is perhaps the strangest origin story ever.

According to one version of the tale, Pegasus arose from a battle between the hero Perseus and a vicious sea monster, Cetus. During Perseus's battle with Cetus, Medusa's head (which Perseus had cut off to turn the sea monster into stone) was still bleeding. When Perseus held out Medusa's head to the sea monster, some blood dripped out of her neck. When the blood hit the seawater a magical thing happened: Medusa's blood mixed with the water and turned into Pegasus, the Flying Horse. This was actually a common theme in Greek mythology—when a monster or god bled, something always sprang from it. Still, it is hard to imagine a lovely winged horse coming from such a gruesome beginning.

If that wasn't confusing enough, from Greece the constellation Pegasus seems to fly upside down in the sky (no explanation for this was given). The good news is that unlike other star patterns in the Southern Hemisphere that seem upside down, from the mid-southern latitudes Pegasus actually flies right-side up.

Look for a big square of four semi-bright stars. This is Pegasus's body. His head, mane, and front legs jut out to the left of the square. You may imagine that you can see Pegasus's back legs toward the right, but those stars are actually part of Andromeda. The four stars making up the Great Square of Pegasus are interesting sights in their own right. Starting from Alpheratz (also Andromeda's head) and heading clockwise around the square, you will come to Scheat, then Markab, and then Algenib. The names of these stars are Arabic in origin and are meant to help illustrate different body parts of the horse. Scheat means "shoulder" or "upper arm," Markab means "saddle," and Algenib means "the side" or "the wing."

The four stars in this square are all different colors and distances away. Alpheratz is blue-white in color and lies about 97 light-years from Earth; Scheat is a red star 196 light-years away; Markab is bluish-white and 133 light-years out; and Algenib is a deeper blue and a whopping 390 light-years away.

How to Find It

Pegasus is the constellation of spring for the Southern Hemisphere. The best way to find it in the sky is to look toward the north on spring evenings and look for the Great Square of four stars marking its body. By the end of September the four stars will have risen above the northeastern horizon and look more like a diamond than a square, with the star Alpheratz at the bottom and Markab at the top. In October Pegasus is about 30 degrees up in the northeast and by November the flying horse soars to its highest point, due north with an altitude of about 40 degrees. By December it is lower in the northwest and by New Year is barely visible above the northwestern horizon.

You may also recognize the Great Square of Pegasus by noting a lack of stars within the square. The Great Square contains surprisingly few naked-eye stars for such a large area of the sky.

ANDROMEDA, THE PRINCESS

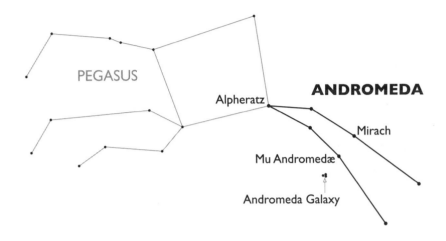

What Is It?
Constellation

Difficulty Level
Moderate

Description
In Greek mythology Queen Cassiopeia bragged about her beauty all the time, but one day she offended the god of the sea, Poseidon. In this version of the story she said that her daughter, Andromeda, was more beautiful than all the mermaids in the ocean. When Poseidon heard about this he was outraged and stormed to the castle to confront the king and queen. "Being a fair and angry god," Poseidon said, "I will give you two choices for your punishment: I will send down a tremendous tidal wave on your land killing everyone and everything or I will take your one and only daughter, Andromeda, chain her to the big rock in the sea, and let my sea monster eat her."

Cepheus and Cassiopeia didn't have much choice. Either way their daughter was doomed. If the king selected wisely, he could at least save his kingdom from total destruction. Reluctantly, the queen's men rowed

Andromeda out to sea, chained her to the sacrificial rock in the ocean, and waited. Poseidon's sea monster then emerged from the briny deep and was about to eat Andromeda. Who arrives just in time to save her? Perseus, the hero! Perseus aimed the head of Medusa at the sea monster and the monster immediately turned to stone. Andromeda was saved!

Alpheratz is a large blue-white star that lies about 97 light-years from Earth. The name comes from Arabic, meaning "the navel of the mare." And this is where it gets complicated. You will notice that not only does Alpheratz mark the princess's head but it is also one corner of the Great Square, the stars marking the body of Pegasus, the Flying Horse (see illustration in the Pegasus, the Flying Horse entry). So Andromeda's head doubles as Pegasus's rear. It is not a glamorous place for a princess, but that means Alpheratz will help you identify two constellations in the autumn sky.

How to Find It

To find Andromeda, you need to find the star Alpheratz. First locate a large square of four semi-bright stars. This is the Great Square of Pegasus that marks the body of the stellar flying horse. Alpheratz is the star at the bottom right corner of the square. This is Andromeda's head. Andromeda's stars form a skinny letter "A" shape with Alpheratz at the top. Her body and legs point down and to the right with the right side of her body noticeably brighter than her left side. You can find Alpheratz low in the northeastern sky in October, with the rest of Andromeda's stars emerging by November. In December face north and look for her head and legs just above the northern horizon, and by January Andromeda seems to lie flat just above the northwestern horizon. Some stars in Andromeda may be too low in the sky to see well from mid-southern latitudes, so having a clear view to the northern horizon will greatly help you find the chained princess.

ANDROMEDA GALAXY

What Is It?
Galaxy

Difficulty Level
Difficult

Description
We now come to the farthest thing you can see with the naked eye: the Andromeda Galaxy. Also known as M31, the galaxy is composed of approximately 1 trillion stars that lie about 2.5 million light-years away.

Stargazers throughout the Northern Hemisphere have noticed the Andromeda Galaxy for millennia without fully knowing what it was. In A.D. 964 Persian astronomer Abd al-Rahman al-Sufi observed the Andromeda Galaxy and called it a "little cloud" in his *Book of Fixed Stars*. This is probably the best description of what you can see of M31 on a good night.

Of course, you can more easily spot M31 with a pair of binoculars. But even if you observe the Andromeda Galaxy through a telescope, you'll still see a fuzzy, cigar-shaped blur with a brighter central core. The view might not impress you at first, but remember that you are seeing one trillion stars shining at you from more than 23 quintillion kilometers away.

How to Find It
Viewing the Andromeda Galaxy from the mid-southern latitudes is definitely a challenge. You must have a clear view to the northern horizon and be in an extremely dark sky, free from light pollution. Furthermore, there is really only one favorable month to locate this galaxy when it is high enough in the sky: November. Face north and locate Andromeda's head star, Alpheratz, at the bottom right corner of the Great Square of Pegasus. Then follow along her body until you reach her hip stars, the brighter one named Mirach and the dimmer one called Mu Andromedae. Connect the dots from Mirach to Mu Andromedae, and then continue that line for another 3–4 degrees. You will run right into the Andromeda Galaxy. The distance between Mirach and Mu, and Mu and M31 are about equal, so the hip stars will make you hip to the galaxy's location!

PERSEUS, THE HERO

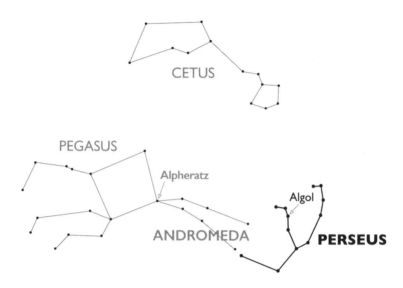

What Is It?
Constellation

Difficulty Level
Difficult

Description
Perseus was the son of the Greek god Zeus and a mortal woman named Danaë. Favored by the gods, Perseus received some valuable gifts. From Hephaestus, the blacksmith god, he got a sword that would cut through anything; from Athena, the goddess of wisdom and warcraft, he received a shield that was shiny as a mirror; and from Hermes, the messenger god, he was given a pair of winged sandals so he could fly through the air with ease.

Perseus heard about how Andromeda had been chained to a rock to be sacrificed to Cetus, the sea monster, and he vowed to save her. He first went to the cave of Medusa (the Gorgon who had snakes for hair and turned you to stone with just one look). Medusa's reflection, although still ugly, wouldn't turn you to stone, so Perseus used Athena's mirror-like shield to warn him of Medusa's approach. When she got close enough, Perseus closed his eyes and cut her head off with Hephaestus's sword. He put the severed head in a bag and flew on his winged sandals to Andromeda's rescue.

Perseus flew as fast as he could and made it just in time to see the fair Princess Andromeda about to be devoured by the sea monster. He yelled down to Andromeda to close her eyes, then he closed his eyes tight and pulled out Medusa's bloody, severed head. He showed the head to the sea monster, who instantly turned to stone and fell to the bottom of the ocean, never to be seen again. Perseus unchained Andromeda, and they lived happily ever after.

The constellation Perseus is much more indistinct than his daring story might suggest. His outline is a squiggle of semi-bright stars that do not stand out on their own and are barely visible from the Southern Hemisphere.

How to Find It

The constellation Perseus lies so far north in the sky that it barely flies above the northern horizon during spring evenings. The best, and nearly the only, month to see Perseus in the evening sky is between November 15 and December 15 when his outline is just above the treetops. First, turn to the northwest and locate the Great Square of Pegasus. The star on the bottom right is Alpheratz, which is also the head of Andromeda. Follow the fainter stars of Andromeda to the right, and 40 degrees from Alpheratz you will run into the body of stars that is Perseus. If it helps, you can picture Perseus at Andromeda's feet.

One other landmark in the sky can help you find Perseus: the Pleiades. The little cluster of stars also known as the Seven Sisters will hang about 20 degrees above the stars of Perseus in the early December evening sky.

ALGOL, THE GHOUL

What Is It?
Star

Difficulty Level
Difficult

Description
To save Andromeda, Perseus showed Medusa's bloody head to the sea monster and turned him to stone. The star in the Perseus constellation named Algol marks the pivotal point in this epic battle. It is the star that was said to be the head of Medusa. And it wasn't just the Greeks who thought that this was a monstrous star. Ancient stargazers from around the world noticed something strange about Algol. Sometimes Algol was bright, and at other times it was dim. Whatever was happening to it, these changes were interpreted as bad omens. All the other stars seemed so steady and dependable that Algol's fluctuation of light scared ancient astronomers around the world. The word *ghoul* comes from the Arabic name for this star. Other cultures referred to Algol by such colorful names as "Satan's Head," "Demon Star," and "Piled-Up Corpses."

In reality Algol is two stars that revolve around each other. Earth just happens to be in the prime position to see the stars eclipse each other at regular intervals. When one star eclipses the other, a lot of light pointed toward Earth is blocked out. You can see a noticeable dip in starlight about every 2.86 days.

How to Find It
Picture Perseus lying on his back just above the northern horizon in December. Algol (as Medusa's head) is the semi-bright star at the end of Perseus's upraised arm. He is holding it above him and pointing it at the head of the sea monster constellation Cetus that, as you'll learn in the following entry, swims in the sky high above. It can be a challenge to find Algol when it dims, so if you detect a change in the outline of Perseus from night to night and one star seems to be missing, the culprit is the star Algol.

CETUS, THE SEA MONSTER

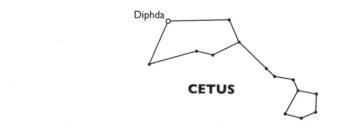

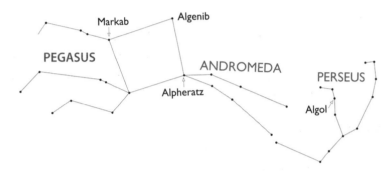

What Is It?

Constellation

Difficulty Level

Moderate

Description

In Greek mythology Cetus represents the sea monster that the god of the sea, Poseidon, could dispatch whenever and wherever he wanted individual humans to feel his wrath. In this case Poseidon sent Cetus to consume the fair maiden Andromeda in order to punish her family's vanity.

Andromeda was chained to the rock in the sea and soon Cetus emerged from the briny deep. Just as Cetus was about to chow down, who came flying along on his winged sandals? Perseus, to the rescue! When he arrived on the scene, he looked down to see the maiden in distress. Without a moment to lose, he yelled, "Andromeda! Close your eyes!" Perseus closed his eyes, reached in his bag, and pulled out Medusa's bloody, severed head and showed it to the sea monster. Cetus took one look at it and instantly turned into stone, cracked of his own monstrous weight, and sunk to the bottom of the ocean, never to be seen again. Perseus unchained Andromeda, they fell instantly and madly in love, and the couple flew away to live happily ever after.

Cetus marks the easternmost point of a watery realm in the heavens. The ancient Greeks designated several watery constellations in the spring sky, including Piscis Australis, the Southern Fish, along with several faint and difficult-to-find constellations like Aquarius, the Water Bearer; Pisces, the Fish; and Capricornus, the Sea Goat. These join other water-loving winter constellations described earlier—Delphinus, the Dolphin and Cygnus, the Swan.

How to Find It

Set apart from the rest of the constellations in the spring sky, Cetus, the Sea Monster can be a challenge to locate even though it is a large constellation about 40 degrees long. It has several semi-bright stars, but only one, named Diphda, really stands out. If that wasn't bad enough, the outline of his stars resemble an upside-down recliner chair with a lumpy headrest more than they do a vicious sea monster. To locate him, first find the Great Square of Pegasus. Connect the dots between the stars Alpheratz and Algenib on the right side of the square and travel upward another 30 degrees. There you will find Cetus's brightest star Diphda and the tail of the sea monster. To find Cetus's head, connect a line from the stars on the top of the Great Square, Markab and Algenib, and continue almost another 40 degrees.

You can start to see Cetus rising in the east after sunset in September and October. In November and December, it hovers high up in the north about 60 degrees above the horizon. By January Cetus is about halfway up in the northwest, and by February it starts to set in the west after dark.

PISCIS AUSTRALIS, THE SOUTHERN FISH

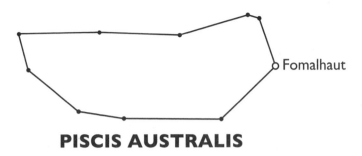

Fomalhaut

PISCIS AUSTRALIS

What Is It?
Constellation

Difficulty Level
Moderate

Description
Piscis Australis (or Piscis Austrinus) is a little constellation that swims up to the roof of the heavens during the middle of the spring. Piscis Australis is the Southern Fish, and the big daddy of the two fish that form the zodiac constellation Pisces.

Piscis Australis has one really bright star called Fomalhaut (pronounced Foam-a-lot), which means "fish's mouth." The other stars in the constellation are so faint that none of them have common names. The outline of these fainter stars is often drawn in an oval shape.

How to Find It
The best way to locate Piscis Australis is to find Fomalhaut, its brightest star. It starts rising in the southeastern sky in July and then can be found high in the east during August and September. During October it's best to lie flat on the ground to see this fish in the sky since it will be straight overhead, with bright Fomalhaut on the right side of the star pattern. Then November and December Fomalhaut is high in the western sky but you can still catch it through January in the southwest.

FOMALHAUT, THE FISH'S MOUTH

What Is It?
Star

Difficulty Level
Easy

Description
During the spring evenings one star stands alone high in the sky: Fomalhaut. One translation of this star's name is "fish's mouth." But this star was also known to some ancient astronomers as the "first frog" that hopped across the sky ahead of the "second frog," the star Diphda, in the constellation Cetus.

There are no other first magnitude stars equal to or brighter than Fomalhaut within 50 degrees of it. Its solitary location made a perfect landmark for measuring the spring sky. In addition, Fomalhaut is one of the closer bright stars to us at only 25 light-years away. Its proximity has made it a great target for astronomers. This star is so bright that astronomers decided to mask out its glare in order to see what's around it. When they did this, they found rings and rings of dust circling around Fomalhaut. Embedded in that dust was a planet named Dagon, looking like a slow-moving lump around the fish's mouth. This was the first planet to be seen visually orbiting another star.

Some astronomers say that there aren't any green-colored stars, but this is open to individual interpretation. Fomalhaut twinkles blue, white, and sometimes green. Find Fomalhaut for yourself and see what you think.

How to Find It
Fomalhaut, the lone star of the spring, is the brightest star in the constellation Piscis Australis. Look for it rising in the southeast in July, standing very high in the east in August and September, and then reaching the zenith in October. In November and December it is still high up but in the west, and it will set in the west after January.

GRUS, THE CRANE

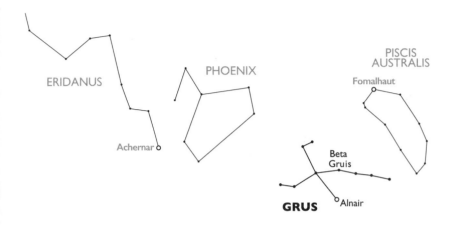

What Is It?
Constellation

Difficulty Level
Moderate

Description
How does a constellation become official? Well, in the case of Grus, the Crane, it took a while for people to accept it. In the late 1500s a pair of Dutch navigators, Pieter Dirkszoon Keyser and Frederick de Houtman, took a trip to the Southern Hemisphere and came back to Europe with detailed star charts of the southern sky. These stars that were not visible from Europe excited the imagination of astronomer and mapmaker Petrus Plancius, who added them to his star charts. He basically invented the constellation Grus, the Crane.

Soon after, another astronomer looked at the same stars and formed them into a flamingo constellation. And in ancient Arabia, some of these stars were linked to the constellation Piscis Australis, the Southern Fish. Which version was best? None of them had a strong mythological story to accompany the star patterns, but in the end the Crane was more popular and survived to this day.

What do you see in these stars? Can you see a crane, a flamingo, or an extra tail for the Southern Fish? Its outline looks like yet another cross shape, but Grus is much more subtle than the Southern Cross or even the False Cross. Most of the stars in Grus are about third magnitude and not easily seen in city skies, but one bright star will catch your attention: Alnair, which appropriately means "the bright one." Alnair is a first magnitude star, is blue-white in color, and lies about 100 light-years from Earth. Beta Gruis, which lies about 177 light-years away from Earth, is the star just to Alnair's left and is nearly as bright. Beta Gruis is a red giant star.

How to Find It

Grus, the Crane is located so far south in the heavens that it is nearly circumpolar. While you can find it somewhere in the southern sky almost all year round, it appears highest above the southern horizon during the spring evenings when it is nearly straight overhead. After dark on most spring nights you can find the crane below Piscis Australis and to the right of both the constellations Phoenix and Eridanus. When you look very high in the south after dark in the spring notice the three brightest stars forming a huge triangle. They are Achernar (in Eridanus), brightest and to the left; Fomalhaut (in Piscis Australis) up above; and faintest of the three, Alnair (in Grus) on the lower right corner.

You can start to see Alnair and the rest of the constellation rising in the southeastern sky in July. By August and September these stars are about halfway up in the southeast. In October and November you will have to lie back on the ground with your feet pointing south to see the crane well since it will be about 75 degrees up in the sky from mid-southern latitudes. In December Grus is located about halfway up in the southwestern sky. By January evenings it appears very low in the southwest and disappears from view at the beginning of February.

ARIES, THE RAM

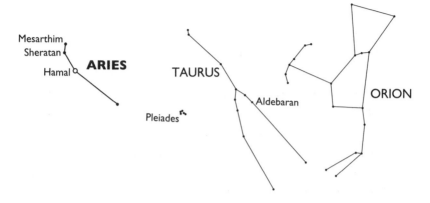

What Is It?
Constellation

Difficulty Level
Moderate

Description
For such a tiny constellation Aries features mightily in several Greek myths. One begins when a king, with a son named Phrixus and a daughter named Helle, leaves his wife and remarries a mean woman. The new wife is jealous of the children and plans to sacrifice them to the gods. At the last moment their biological mother sends a magical, winged, golden ram to rescue Phrixus and Helle and fly them away to the east. Unfortunately, the daughter falls off the flying ram to her death in an area still called the Hellespont (named in her honor).

Phrixus lands safely, sacrifices the ram to Zeus, and gives the Golden Fleece to another king named Aeetes (whose daughter Phrixus eventually marries).

Years later a Greek hero named Jason gathers a group of explorers. Calling themselves the Argonauts, they set out to reclaim the Golden Fleece. (Their ship, the *Argo*, was turned into a large constellation—see the entry for Carina in the section on the Autumn Sky.) The Argonauts, which included Hercules, Orpheus, Castor, and Pollux, among others, have a series of epic adventures and ultimately succeed in their quest for the Golden Fleece.

The stars of Aries don't look anything like a ram, however. The outline of the three major stars in Aries resembles the slight curve of a boomerang. Only one star in the constellation is significantly bright: Hamal. The others are second, third, and fourth magnitude stars that are difficult or impossible to see from skies with light pollution.

Hamal lies 66 light-years away and is orange in color. Its name is an Arabic word meaning "head of the sheep." The star has also had many other cultural names, such as "the Ram's Eye" and "the Horn Star." And finally, eight temples of ancient Greece were dedicated to this star—the Greeks do value their sheep.

How to Find It

Aries is extremely difficult to find because it has only three stars of note: Hamal, Sheratan, and Mesarthim. But other stars and constellations can help you locate the ram's place in the sky. First find Orion's Belt. Then connect the dots of the belt and continue that line to the left or east for 23 degrees. This view will take you under the ruddy star Aldebaran, which is in Taurus, the Bull. Keep going another 12 degrees until you reach a tight, bright group of stars called the Pleiades. Continue that line, arcing it a little upward, and after another 23 degrees you will run into Aries's brighter star, Hamal.

Aries first pops up in the northeastern sky during November. By December the ram is visible when you face due north and look about halfway up in the sky. In January it is only about a third of the way up in the northwestern sky and by February will set soon after dark.

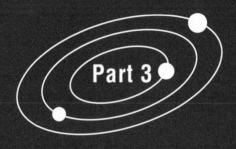

Part 3

BEYOND STARGAZING

So far we've discussed how to view the Sun safely, what to look for on the Moon, the motions of the five naked-eye planets, and how to identify dozens of stars, constellations, and several deep space objects. What else can you see with the naked eye on a clear night? Lots.

If you spend any time stargazing, every so often you'll definitely notice a slow-moving, nonblinking light crossing the heavens. What you're seeing is one of the 4,000 man-made satellites that circle the Earth every night. Launched into space by rockets, these satellites aid global communications, conduct scientific research, and sometimes undertake secret missions. Satellites shine only from reflected sunlight. While you are standing in the dark on the ground, the satellites are high enough to be lit by the Sun. The sunlight reflects off their surfaces, and you can see them against the darker background, making them look like little stars sailing through the blackness of space.

Or, on any given night you might also see a meteor streaking across the sky. Most people call them shooting stars, but their origins and fiery deaths are a lot closer to home than the distant stars. Meteors are best seen during annual meteor showers, when you can see several dozen crossing the sky per hour. The greatest showers, which occur once in a generation, are called meteor storms. You can see so many shooting stars pass overhead that it may look like the sky is actually falling.

Finally, you can observe some of the greatest astronomical events with the naked eye. Some of these sights are quiet and subtle, like observing the Milky Way or zodiacal light. Some require timing and patience, like observing planetary conjunctions and occultations. And others, like auroras and eclipses, will blow your mind. You have to be at the right place at the right time (and perhaps have a little luck), but these special heavenly events will wow you.

Catching Satellites

It's not a bird, not a plane, and not a meteor. That steadily moving light in the sky is probably a satellite. How do you know for sure? There are some surefire ways to tell. If you see a blinking light in the sky, it is probably a plane. Satellites almost always shine with a steady light. They may gradually brighten or dim, but they do not blink.

Satellites also never suddenly change direction. They always travel in a steady arc across the sky. They are fast but not flashy, and the speed of a satellite depends on its altitude above Earth. Most satellites take 4–6 minutes to cross the sky from horizon to horizon. If you see something moving much more quickly (if it crosses the entire sky in only a few seconds), know that you probably saw a meteor/shooting star.

The brightest satellites and the easiest to spot are the International Space Station (ISS), Tiangong-2 (a prototype of a future Chinese space station), and various iridium satellites. With a little practice you might be able to find the X-37B (a supersecret unmanned space plane also known as an Orbital Test Vehicle, or OTV).

To know when to look for certain satellites, websites like www .heavens-above.com and apps like Satellite Tracker can help. Once these tools know your location, they can tell you the exact time when, for example, the ISS will pass overhead, in which direction you should look, how high in the sky it will fly, and how bright it will appear. With this prior knowledge you can amaze your friends by taking them outside and showing them a satellite cutting across the sky in the exact position that you predicted. They'll think you're a rocket scientist!

THE INTERNATIONAL SPACE STATION

What Is It?

Space station (manned satellite)

Difficulty Level

Easy

Description

What is 108 meters long, 72 meters wide, weighs nearly 400,000 kilograms, and circles Earth every 92 minutes? It is the International Space Station (ISS), the largest satellite humans have ever constructed in space.

The ISS is a joint effort between the United States, Russia, and several other countries. It went up into space, piece by piece, starting in 1998. It is so large that it can accommodate a crew of six full-time residents as well as entertain visiting guests (the replacement crew). The purpose of the ISS is to provide a laboratory in space to test out new technologies that can help us better understand Earth and prepare us for a longer space mission to Mars and beyond.

Like most satellites the ISS is surprisingly close to Earth. It orbits between 330 and 410 kilometers up, which means you could drive there in about three hours (if it weren't straight up). To stay in orbit, the ISS must travel at more than 27,000 kilometers per hour. Every once in a while rocket boosters bump it into a higher orbit so that this object the size of a football field will not fall to Earth.

Astronauts aboard the ISS get an amazing view of Earth and the sky. Since they circle Earth so quickly, they experience one sunrise and one sunset about every 90 minutes.

How to Find It

From Earth the ISS looks like an exceptionally bright, slowly moving star. Like other satellites (and unlike stars), it should not blink. If you were to observe its entire path across the sky, it would take about 5–6 minutes for it to travel from horizon to horizon.

Because of its massive size the ISS can shine like a dazzlingly bright beacon. Even at its minimum brightness, it shines with a glow that is similar to a first magnitude star. When it flies directly overhead or reflects the maximum amount of sunshine toward you, the ISS can be almost as bright as the planet Venus. The ISS can fly over your town from nearly any direction and at nearly any time. Satellite tracking apps and websites can alert you to when and where to look. When you see it, be sure to wave at the astronauts inside who are about 350 kilometers above you.

TIANGONG-2

What Is It?
Space station

Difficulty Level
Moderate

Description
China is a major player in the space race and has launched several manned missions into orbit. Recently they created and briefly occupied their own space station. It was called Tiangong-1 and it held a crew of three people for a few days in 2012 and 2013. It served as a testing ground for future space missions and outlived its two-year mission plan. But in 2016, Tiangong-1 stopped responding to controllers on the ground and began to drift out of control. In 2018 Tiangong-1 crashed back down to Earth, largely burning up as it plunged through the atmosphere.

Enter onto the scene Tiangong-2, another Chinese space station that has been circling the globe since 2016. It's 10.4 meters long and 3.35 meters wide and weighs 8.6 tonnes. The accommodations aboard Tiangong-2 are quite cramped compared to the massive International Space Station, but two astronauts lived there for 30 days. They conducted psychological and scientific tests and released a satellite into orbit. The two-man crew then left Tiangong-2 and returned to Earth on November 18, 2016. Although no one has revisited this Chinese space station, it continues to circle the globe and can be seen occasionally from Earth with the naked eye.

How to Find It
A larger Chinese space station is scheduled to launch in 2020. Until then you can look for Tiangong-2 flying over your town. It can be about as bright as a first magnitude star like Spica or Regulus. Like all satellites, Tiangong-2 will not blink. Look for a steady light traveling at a speed that would take it from horizon to horizon in about 5 minutes. Like the ISS, Tiangong-2 could fly overhead from nearly any direction at nearly any time. Use satellite tracking apps and websites to help you predict the best times and directions to spot it.

IRIDIUM SATELLITES

What Is It?

Satellite

Difficulty Level

Easy

Description

The brightest man-made objects you can see in the sky are called iridium satellites. These sixty-six communications satellites have highly reflective exteriors, so when the sunlight hits them squarely, a beam of light flashes toward your eye for a brief moment. Amateur astronomers call these flashes iridium flares, and they can be dozens of times brighter than the planet Venus. If you are taken by surprise by an iridium flare, it can be quite disconcerting. Because they occur only for a short period of time (usually around 5–10 seconds), you might think a rocket just exploded, a meteor fell to Earth, or that the alien invasion has started. But if you are expecting to see an iridium flare, it is a really cool event to witness.

How to Find It

With iridium flares you have to be ready. Fortunately, websites like www.heavens-above.com and apps like Sputnik and Satellite Tracker can predict when and where to look for these bright flashes. Make sure the app or website knows your specific location. The brightest flashes occur over localized areas so you have to be in the correct location at the right time. They are best seen right after sunset and right before sunrise, but iridium flares can be so bright that you can occasionally catch them during the daytime.

X-37B OR OTV

What Is It?
Spacecraft

Difficulty Level
Difficult

Description
You may be surprised to hear that a secret, unmanned, reusable US Air Force spacecraft, an Orbital Test Vehicle (OTV) nicknamed the X-37B, circles the globe from time to time. It was first observed by amateur astronomers in 2010, when no one other than the US military knew about it. What is the X-37B doing? The US Air Force gives very few details. "Tests," they say. "What kind of tests?" you may ask. The Air Force is mum, so no one is really sure.

The X-37B is launched into space by a rocket, discards the rocket, and then circles Earth. After months or even years, it then reenters Earth's atmosphere and lands like a plane at Vandenberg Air Force Base in California or at NASA's Kennedy Space Center in Florida. A version of this spacecraft/satellite has completed four missions of various lengths. Orbital Test Vehicle 4, or OTV 4, was in orbit between May 20, 2015, and May 7, 2017. OTV 5 launched in September 2017.

How to Find It
Whenever the X37-B flies, you can be sure that amateur astronomers will continue to keep tabs on it and publish its path on websites like Heavens Above. So even if you don't know exactly what this satellite is doing, you can know where and when it soars overhead. The X-37B can shine with the equivalent brightness of a first, second, or third magnitude star. At its brightest it is as bright as the stars Vega and Acrux, but when the Sun's angle is not optimal, it can appear fainter than the stars in Orion's Belt.

Meteors and Meteor Showers

Head outside on any clear night, sit back in a comfortable chair with a warm beverage, and watch the light show up above. Chances are, if you sit out there long enough, you will see a meteor streak across the sky. Also called shooting stars, meteors are objects falling through Earth's atmosphere.

A meteor exists for only a brief moment; most are only visible for a few seconds. And, despite their occasional brilliance, most meteors are incredibly small. They are usually about the size of a grain of sand and rapidly burn up before hitting the ground.

When you see meteors, they may look close to you but they are not. They light up when they are about 60–80 kilometers above Earth in the upper atmosphere. Meteors blaze because they are decelerating from tens of thousands of kilometers per hour to hundreds of kilometers per hour. That rapid deceleration transmits into heat and causes the air around the meteor to glow. Once the meteor is lower in the sky and has stopped this rapid deceleration, it stops shining.

Meteor showers occur at predictable dates each year when Earth slams into a swarm of space debris left behind by passing comets or asteroids. To maximize your meteor shower viewing experience, get out of the cities and away from lights. The darker the sky, the more meteors you will see. The best times to view meteor showers are generally between 2 a.m. and 5 a.m. You can see some early shooting stars around midnight, but the later you stay up the better. Once you start to detect hints of sunrise coming in the east, the number of meteors you'll see will greatly diminish. So let's take a look at some shooting stars!

FIREBALLS, GREAT BALLS OF FIRE

What Is It?
Meteor

Difficulty Level
Moderate

Description
Really bright meteors are called fireballs. They can glow all sorts of colors including white, blue, and green. Really dramatic fireballs can break up into multiple pieces and become multiple meteors streaking across the sky.

Most fireballs were originally pieces of asteroid fragments. Astronomers have charted more than 600,000 asteroids in our solar system. These irregularly shaped chunks of metal, rock, and dust circle our Sun mainly between the orbits of Mars and Jupiter. But some very small ones can come close to Earth and even run into us.

And that is exactly what happened on February 15, 2013, when a huge meteor streaked across the sky over Chelyabinsk, Russia. For a brief moment it shone brighter than the Sun and cast stark shadows. At first it did not make a sound, but about 2 minutes later a sonic boom shattered windows and even knocked people over. The speed of light is 300,000 kilometers per second while the speed of sound is only 1,235 kilometers per hour. That difference in speed accounted for the lag time between seeing it and hearing it. If you see an incredibly bright fireball explode above the ground, stay away from windows. A sonic boom may strike a few minutes later and blow out the window in front of you. However, not a single person in the past one hundred years has been struck and killed by a meteorite or falling space debris, so there is nothing to fear when you are conducting your meteor watch.

How to Find It
Every day between 9 and 90 tonnes of material falls into the atmosphere from outer space. While you can sometimes see a chunk of man-made debris entering the atmosphere and turning into a fireball, the vast majority of it burns up and/or falls into the ocean. Observing a fireball is a rare and random event that astronomers cannot yet predict.

THE PERSEIDS, ORIONIDS, LEONIDS, AND GEMINIDS

What Is It?
Meteor showers

Difficulty Level
Moderate

Description
After centuries of watching the skies, astronomers figured out that more meteors fall on certain days of the year than on others. These events are called annual meteor showers and several of them provide consistently impressive celestial performances.

When a comet passes near the Sun, it sheds icy material to form a long tail. Tail particles are mainly tiny pieces of ice and dust that can accumulate in great clouds of debris along the comet's orbit. Earth slams into this cometary debris on a yearly basis, and the result is a predictable schedule of meteor showers. And when these clouds intersect with Earth, get ready.

The most impressive annual meteor showers are the Perseids, Orionids, Leonids, and Geminids. Meteor showers get their names from the constellation from which the shooting stars seem to radiate. The meteors don't actually fall from the stars in Perseus, Orion, Leo, and Gemini—they just appear to do that. They actually ignite in Earth's atmosphere, about 60 kilometers up.

Astronomers know exactly where these meteors come from. The Perseids are remnants of the comet Swift-Tuttle, the Orionids come from Halley's Comet, the Leonids stream from the tail of comet Tempel-Tuttle, and the Geminids originate from pieces broken off an asteroid named 3200 Phaethon.

The best times to view meteor showers are generally between 2 a.m. and 5 a.m. You can see some early shooting stars around midnight, but the later you stay up the better. Don't use telescopes or binoculars; your naked eye is all you need. Take in as much of the sky as possible to catch as many stray streaks as you can.

Get away from the lights of the city and find a dark sky location where you can detect even the faintest meteors. Avoid meteor showers that peak during a Full Moon, because the moonlight will wash out all but the brightest meteors. And keep your expectations low. While some astronomers predict that you'll see dozens to hundreds of meteors per hour, it is more realistic to expect 10–20 per hour.

How to Find It

The Perseids are the meteor shower of winter, peaking each year on August 12 or 13. You want to generally face in the direction of the constellation Perseus, which rises in the northeastern sky at midnight and shifts to the north later at night, but meteors can streak from any direction at any time between midnight and dawn. Be patient and vigilant.

The most famous comet in the sky, Halley's Comet, is responsible for the Orionid meteor shower. The Orionids peak every year around October 21. Although you can't see Halley's Comet in all its glory until it passes near Earth again in 2061, you can still see its remnants from previous flybys in the sky every October. Face the constellation Orion on October 21 between 1 a.m. and 5 a.m. to see the maximum number of Orionids.

The Leonids produce reliably good meteor showers every November 17 or 18. To find them, lie back and search the skies around the constellation Leo as it comes up in the eastern sky after midnight and slowly arcs to the north later at night. If you watch all night until the break of dawn, you may see dozens of meteors per hour.

One annual meteor shower that has become increasingly visible in the twenty-first century is the Geminids. Every December 13 or 14, scan the sky around the constellation Gemini between 2 a.m. and 5 a.m. to see the peak of this unusual meteor shower. The Geminids are not cometary in origin—they actually come from an asteroid called 3200 Phaethon, a rocky and icy body circling the Sun. As it regularly flies close to Earth, 3200 Phaethon leaves bits of rock and ice behind. Every December, Earth flies through these asteroid parts and stargazers are treated to a solid meteor shower.

METEOR STORM

What Is It?

Mother of all meteor showers

Difficulty Level

Difficult

Description

Better than a shooting star, better than a meteor shower, a meteor storm is a once-in-a-lifetime experience. Like meteor showers, meteor storms originate from the leftover debris of comets. While during normal meteor showers Earth plows through some of this debris, during a meteor storm our planet crosses paths with swarms of denser cosmic material. During a meteor storm you can observe about one hundred meteors per minute raining down from all directions. This is the heavenly fireworks show that made witnesses, both ancient and modern, truly think the sky was falling.

The annual Leonid meteor showers have produced the most memorable storms in recent history. In a fiery display in November 2001, observers counted more than 2,000 Leonid meteors per hour. But that was nothing compared to the Leonid meteor storm of 1833 where some astronomers estimated between 100,000 and 240,000 meteors streaked across the sky in one night. Astronomy writer Agnes Clerke described the scene: "On the night of November 12–13, 1833, a tempest of falling stars broke over the Earth....The sky was scored in every direction with shining tracks and illuminated with majestic fireballs." Another witness to the stunning scene, John Sharp, expressed the effect this meteor storm had on the townspeople: "The Meteors fell from the elements the 12 of November 1833 on Thursday in Washington. It frightened the people half to death." If you are not expecting pieces of the heavens showering down from the skies above you, a meteor storm can be downright terrifying. But remember, none of these "shooting stars" has killed anyone in the past one hundred years and they are just awesome to witness.

How to Find It

Unfortunately, meteor showers rarely turn into meteor storms. Astronomers have linked the Leonids to the debris from comet Tempel-Tuttle, which circles the Sun every thirty-three years. And almost every thirty-three years Earth slams into thicker remnants from this comet's tail. Since we know Earth's orbit and Tempel-Tuttle's orbit in great detail, astronomers try to predict which years could put on a better meteor show than others. After a superb display in 1966, astronomers figured that the next possible meteor storm would occur in 1999. It didn't happen that year. Instead, the big storm came two years later, in 2001.

That's the thing with meteors: they are predictably unpredictable. Assume nothing when you try to look for shooting stars. The key is to observe the sky between 2 a.m. and 5 a.m. during the peak of each meteor shower. You will most likely see just a few, but if you're lucky and a meteor storm breaks out, don't panic. Just sit back, "Ooh" and "Aah," and enjoy the grand show.

Additional Astronomical Events to See with the Naked Eye

There is a loose organization of people from all walks of life who make it their purpose in life to see eclipses. They save money and plan vacations around astronomical events. They travel to the far corners of the globe for a glimpse of the greatest shows from Earth. They are called eclipse chasers. Witnessing a total solar eclipse is a powerful experience. When the Moon covers the entire Sun and turns the sky dark in an instant, it does something to people. They yell, they cry, they hoot and holler. Even rational individuals are temporarily transformed into giddy little children under the influence of the total solar eclipse. Eclipse chasers relish this rush, and afterward they have a hard time thinking of anything else except seeing another.

This is the power of the universe. Some astronomical events like total solar eclipses are just downright awesome. And when the northern lights ignite above you in the upper atmosphere or a dynamic comet haunts the night sky, you will remember it for the rest of your life. When the weather is perfect, light pollution is at a minimum, and the heavens align, the universe can put on the most incredible light show. And with a lot of planning and a little luck you can share in the most spectacular, can't-miss, you-gotta-see-to-believe astronomical events in the daytime and nighttime sky. Put these on your bucket list to behold sometime in your life.

In this final section I challenge you to take a trip to the country and get away from the city lights to see the Milky Way and zodiacal light on a moonless night. Prepare for perfect celestial lineups like lunar eclipses, planetary conjunctions, and lunar occultations. Get surprised by the sudden beauty of a great comet or haunted by the appearance of the aurora australis (or southern lights). And chase a total solar eclipse around the world. These final things to see in the sky range from supercool to life changing, and like everything described in this book, you can experience them all with just your naked eye.

PLANETARY CONJUNCTION

What Is It?
Moon and planets

Difficulty Level
Easy

Description
Have you ever seen a bright "star" near the Moon on a dark evening? If you took special note of the scene, you'd see that the "star" was probably a planet. Every month the Moon slides past all five naked-eye planets (Mercury, Venus, Mars, Jupiter, and Saturn). Our solar system is like a flat disc and the planets circle the Sun on nearly the same level plane. From Earth this disc looks like a curved line across the sky, called the ecliptic. Since the Sun and Moon inhabit this line, lunar and solar eclipses can only occur on the ecliptic. The ecliptic cuts through some very famous constellations such as Sagittarius, Leo, Taurus, and Virgo. These constellations, along with eight others, form the twelve zodiac constellations and help astronomers map out the movement of the naked-eye planets. When someone says, "Jupiter is in Leo," they mean that the planet Jupiter can be found in front of Leo's star pattern.

Since you can find all of the planets on or near the ecliptic, they often line up. You may have heard doomsday prophecies like, "Watch out when the planets align!" The truth is, the planets align all the time! That's just what they do.

When a planet and the Moon (or two or more planets) appear exceptionally close to each other, astronomers call that a conjunction. Any combination of the planets and Moon make an impressive sight in the sky. Although two bright lights shining side by side is always an inspiring sight to behold, some conjunctions really make people stand up and take notice.

How to Find It
It is an amazing sight when the brightest planets, Venus and Jupiter, have a conjunction with the Moon or each other. A Venus–Jupiter

conjunction happens about every year but some are more intimate than others. Good conjunctions bring the two planets within one-sixth of a degree. Sometimes they are so close you can barely tell them apart.

The most astronomical planetary conjunction happens when you can see all five naked-eye planets at one time. In May of 2002, Mercury, Venus, Mars, Jupiter, and Saturn were all within 33 degrees of each other. They didn't perfectly line up but instead looked like a planet clump. The next time these five planets will gather so close together will be in 2040, but occasionally you can see them stretched out from horizon to horizon and aligning in the night. To find when the best conjunctions will occur, browse websites that offer "astronomical highlights" for the year ahead. Each month should offer the dates when the Moon cozies up to a planet, and be on the lookout for the rarer occurrence when two to five planets come together in cosmological conjunctions.

LUNAR OCCULTATION

What Is It?
Rare lunar event

Difficulty Level
Difficult

Description
A lunar occultation occurs when the much closer Moon seems to block out the light of a much farther planet or star. It's like a super eclipse. As the Moon orbits Earth it seems to wander across the background stars very quickly (moving about 13 degrees per night). The path the Moon takes is very similar to the ecliptic (the imaginary line where the planets hang out). So the Moon regularly passes through the stars in the zodiac constellations in addition to coming close to the five naked-eye planets. Occasionally, when everything is lined up perfectly the Moon can go in front (or occult) a planet. When the Moon blocks out the light of Venus, Mars, or Jupiter it is especially dramatic since they are typically brighter than an average star.

The Moon occults stars much more frequently than planets. The following are four really bright stars that are close enough to the ecliptic to be occasionally occulted:

- Aldebaran in Taurus
- Regulus in Leo
- Spica in Virgo
- Antares in Scorpius

Even though the Moon passes through the zodiac constellations quickly and repeatedly, it is still very rare for the Moon to occult a bright star or planet. Additionally, these alignments are so precise that they are not visible from everywhere on Earth. You have to be at the right place at the right time to see a lunar occultation.

How to Find It

Every year some astronomy websites list all of the lunar occultations coming up and where you have to be to experience them. Simply search the Internet for "lunar occultations for" and then type in the current year and you'll find a host of solid links.

On the day of a lunar occultation, this is what will happen: When you look at the Moon, there will be a bright star or planet just to the Moon's left. As the minutes pass, you can't see the Moon moving, but the distance between the two objects will noticeably shrink. When it looks like the star is touching the Moon, get ready because in one moment the star will be there—and the next moment it will be gone, covered by the Moon. Now you see it, now you don't. The Moon has no atmosphere, so when the rocky surface of the Moon covers it, the occultation happens almost instantaneously.

When the Moon occults a star or planet, it does so for no more than about an hour. Not only is it amazing to witness the light of a star or planet suddenly disappear, but you can also see it blink back on again as the Moon uncovers it and continues on its journey around Earth.

MILKY WAY

What Is It?
Our galaxy

Difficulty Level
Moderate

Description
For city dwellers the Milky Way is completely invisible. City lights are so bright that light pollution robs millions of people of the chance to see a truly dark starry sky. Because of urban light pollution many people have never seen the Milky Way. But the Milky Way is still there, waiting for you to observe it. And when you see it, it is spectacular!

At first glance the Milky Way looks like a high, thin cloud stretched out in a long line across the sky. It may even seem that your eyes are playing tricks on you. As your eyes adjust to the darkness it looks like someone spilled milk in the heavens. But what you are really seeing is the flat disc of our galaxy. The "milk" is actually billions and billions of stars and the denser part of our galaxy shining at you from tremendous distances. The combined light reaches your eyes as a diffuse cloudy structure—brighter patches where you can discern individual stars interspersed with dark dust lanes that seem to split the Milky Way into channels. Best of all, no telescope is needed to experience it. Just sit back in the grass or in a comfy chair and marvel at your galaxy in all its glory.

How to Find It
It may take a little extra effort (and a small road trip) to see the Milky Way stretch magnificently under a truly dark sky. Drive out of the city on a clear night and you will discover a wealth of stars beyond your imagination. On your next summer vacation make an effort to view the stars from a national park or a really isolated area. Investigate the least light-polluted places in your country and plan a visit. Go there and you'll see how the night sky is supposed to be.

You can see the Milky Way all year round, but some seasons provide better viewing than others. Whenever you see the constellation Crux, the

Milky Way is there too. Alpha and Beta Centauri also seem embedded in a multitude of stars. During the summer months the Milky Way runs near the constellation Orion and between his dogs, Canis Major and Canis Minor. Later in the winter it arches high in the sky near Cygnus, the Swan and Aquila, the Eagle and stretches to Sagittarius and Scorpius.

If you are having trouble finding a dark enough place to view the Milky Way with the naked eye, you can use binoculars to help. After you identify any of the constellations mentioned previously (Crux, Cygnus, Scorpius, etc.), sweep the area in between the brighter stars of these star patterns and you will see a wealth of stars. Just past 20/20 vision the fainter stars of the Milky Way await you.

ZODIACAL LIGHT

What Is It?
Illuminated space dust

Difficulty Level
Difficult

Description
Almost 1,000 years ago the Persian poet Omar Khayyam in his book of poetry, *The Rubaiyat*, wrote his most famous line: "A jug of wine, a loaf of bread—and thou beside me singing in the wilderness." But elsewhere Khayyam made a poetic allusion to a mysterious false dawn. He wrote:

> *Before the phantom of False morning died,*
> *Methought a Voice within the Tavern cried,*

What is a false morning? It took astronomers centuries to figure out what he was writing about, and it makes a rare treat to witness. The event is called the zodiacal light.

Trillions of dust particles populate our inner solar system. These particles are incredibly small and are spread out over millions of square kilometers. But when the Sun strikes them, their combined mass can reflect a diffuse light toward Earth. The reflected sunlight bouncing off the cosmic cloud of comet and asteroid debris creates the zodiacal light that is barely perceptible to the naked eye. But anyone whose eyes have adjusted to the darkness should notice it if they know where to look. This warm, subtle zodiacal light fools late-night stargazers into thinking the dawn is about to break even though daybreak is still hours away.

How to Find It
To see this zodiacal light, you will need to view the sky during or near a New Moon and be far from city lights. If you can see the Milky Way clearly, then you may have a chance to see it. Face east a few hours before sunrise. The zodiacal light will look like a dim, cone-shaped patch of powdered sugar that extends from the horizon to about one-third of the

way up in the sky. It's also called the false dawn because it will give you the illusion that dawn is about to break, even though sunrise is still hours away.

The zodiacal light can also be seen in the evening, a "false dusk," which you can see toward the western horizon a few hours after sunset. Both phenomena are caused by the same thing: trillions of dust particles circling around the Sun and reflecting its light. You can observe both events during any night around the New Moon. However, it is easier to see the morning zodiacal light in the late autumn when the bulk of the particles appear higher in the sky, while spring is the optimal viewing season for the evening zodiacal light when the light reflecting off this space dust is higher above the western horizon.

This extremely subtle light show may not bombard your senses, but once you see it, and realize what you are seeing across millions of kilometers of space, it should evoke a quiet "wow" moment.

A GREAT COMET

What Is It?
A comet of the century

Difficulty Level
Difficult

Description
Comets are small chunks of ice only a few kilometers wide that circle the Sun in long, looping orbits. For most of their lives, they dwell at incredible distances from the Sun and are locked and frozen. But for a few months during every orbit, they may swing so close to the Sun that they heat up. Ices turn to gases, and geysers of material erupt from the nucleus of the comet to form a bright head and a long tail behind it.

About once every decade stargazers are treated to a bright comet that lights up the night sky. The sight of a comet truly inspires us with wonder and awe. There is something about that fuzzy little visitor with the long tail that excites the imagination. Where did the comet come from? Where will it travel to? How long has it been orbiting the Sun? And how many times has it swung by Earth?

There is something unearthly and maybe a little unsettling about a bright comet. It often takes weeks or months to brighten up significantly, so each night you see it, it gets a little bigger. No wonder the ancients feared comets. After all, they looked like they were coming right at them! A comet seems to just hang in the sky with a fuzzy bright head called the coma and long wispy strands of gas jetting off to form a tail. They can haunt the night sky for weeks at a time, brighten unexpectedly, or quickly fade from sight. Even astronomers don't know what they will do.

Although we have had dozens of comets swing by Earth's neighborhood in the twenty-first century, none of them have been truly spectacular. In fact, the last great comet visible to stargazers in the Northern Hemisphere was Comet Hale-Bopp in 1997. This is an extraordinarily long drought. We're overdue for a comet of the century that will wow anyone and everyone who looks up.

How to Find It

A truly great comet is not only visible to the naked eye, but its tail covers 10, 20, or even 30 degrees of the sky. When you walk outside, it immediately grabs your attention; it's difficult to look away. For example, in February 2007, Comet McNaught suddenly and unexpectedly brightened for viewers in the Southern Hemisphere. Previously an underwhelming sight to Northern Hemisphere observers, McNaught seemed to catch fire after it rounded the Sun and headed south. Sporting a tail that seemed to fan out like a peacock, McNaught was the last bright comet seen from Earth. We are definitely overdue!

Unfortunately, astronomers are not expecting any known comets to brightly light up the night in the next few decades. However, most great comets seem to pop up out of nowhere and catch us off guard. They are small, stray ice balls from the outer reaches of the solar system. And we often can't see them until they come closer to the Sun and cast their ethereal tails across the night sky. An optimistic astronomer might say, "Hopefully we will see one next year."

AURORAS

What Is It?
Southern lights

Difficulty Level
Difficult

Description
When you step outside at night and look at the stars, what does it mean if the sky is green? What about if red and white streaks of light start swirling and dancing in waves and curtains? It's not the end of the world. It's the aurora australis, the southern lights!

Pictures do not do this phenomenon justice. Whole swaths of the sky are enveloped in color and light. The weaker shows involve a green-colored fog rolling in from the south while more excited auroras showcase bursts of red and white lines and flares. Although rare, you may see a blue-tinged curtain of light wave and slowly unfurl in the heavens.

Auroras originate from the Sun. Intense solar storms called coronal mass ejections shoot solar material through space at about 2 million kilometers per hour. When some of this material slams into Earth, it excites gases in Earth's upper atmosphere, lighting them up like a neon sign. The solar material can more easily enter Earth's magnetic field around the poles, which makes auroras much more frequent occurrences in Arctic and Antarctic regions.

In the Northern Hemisphere auroras can pop up in Alaska, Canada, Scandinavia, and Russia but can, on occasion, be visible from the mainland United States. These are called the aurora borealis or northern lights.

For the Southern Hemisphere, auroras may be frequent in Antarctica but are much rarer the farther north you live. Nevertheless the southern lights can sometimes be seen in southern South America, Australia, New Zealand, and South Africa.

How to Find It
About once every decade the southern lights can be seen from farther north of Antarctica. When an extremely powerful coronal mass ejection

rockets off the Sun and slams into and bends Earth's magnetic field, it can make auroras visible from Australia, New Zealand, and even South Africa.

When astronomers see extreme bursts of activity they can issue aurora alerts (ways of telling you that southern lights may be visible in your area). Auroras are difficult to accurately predict since there are so many factors going into these light shows. But when a solar storm breaks onto Earth's magnetic field and scatters around the upper atmosphere, particles from the Sun will dance above your head. The good news is that the Sun is 150 million kilometers away, and when astronomers see a large blast coming our way at 2 million kilometers per hour, we still have about seventy-five hours until it reaches Earth. That's a lot of warning! The best place to check for aurora updates online is www.aurora-service .net. This volunteer-run website tracks the southern lights and tells you when and where to go to maybe see them.

For the most part you have to travel south of the 45th parallel to have even a slim chance of seeing auroras. Plan a trip to Antarctica or book passage on an aurora vacation to see the aurora borealis from Alaska, Iceland, or Scandinavia. For inspiration check out www.spaceweather .com to see their aurora image gallery. Pictures do not do the southern lights justice. You really have to experience them.

TOTAL LUNAR ECLIPSE

What Is It?
Alignment of Sun, Earth, and Moon

Difficulty Level
Moderate

Description
There are few astronomical events that demonstrate so beautifully the heavenly dance of the solar system like a lunar eclipse. This rare alignment of the Sun, Moon, and Earth creates one of the best shows in naked-eye astronomy.

A lunar eclipse occurs when the Sun is on one side of us (about 150 million kilometers away), the Moon is on the other side (about 385,000 kilometers away), and the Earth is right in the middle. The Moon circles directly into the darkest shadow cast by Earth into outer space, and our planet blocks the sunlight from reaching the lunar surface. From Earth it looks like a dark shadow is slowly creeping across the face of the Full Moon. When it is completely covered, the Moon turns an eerie shade of pink, orange, or red. Some people call it a Blood Moon, since the darkest lunar eclipses turn the face of the Moon a crimson color. What color will the Moon turn during a total lunar eclipse? Witnessing those slowly evolving colors on the Moon's surface is most of the fun!

How to Find It
To really take in a lunar eclipse, sit outside with friends and family and make a night of it. An average lunar eclipse lasts three hours, and although there are two very exciting parts to it, most of the time you can just relax while the sky show goes on around you.

The best part is capturing the beginning of the eclipse—when the first bit of Earth's shadow can be seen on the Moon. Astronomers predict the start of eclipses down to the exact second, and it can be a thrill to witness the precision of their predictions and see the eclipse first commence.

Over the next hour, Earth's curved shadow will slowly creep across the lunar surface. Bit by bit, the darker shadow covers over craters, seas, and other landmarks.

Then comes the second most dramatic part: totality. When the Moon is completely in the shadow of Earth, it turns an astonishing coppery-orange color. The Moon does not turn completely black since light is still getting to it. Sunlight bends through Earth's atmosphere and weakly bathes the Moon in a warm, orange glow. What you are seeing is the combined light of all the sunsets and sunrises of Earth projected onto the Moon.

Totality can last between 1 and 106 minutes. Be patient and watch how the light and color change during the course of totality. If you look away for a minute and then turn back to the Moon again, you will see a slightly different shade. But all good things must come to an end, and Earth's shadow will eventually pass away and return the Moon to its normal state about an hour after totality. The next total lunar eclipses in the Southern Hemisphere and where they will best be seen are:

- July 27, 2018 (South Africa and Australia)
- January 21, 2019 (South America)
- May 26, 2021 (Australia and New Zealand)
- May 15, 2022 (South America)
- November 8, 2022 (New Zealand and Eastern Australia)
- March 14, 2025 (South America)
- September 7, 2025 (Australia and New Zealand)

TOTAL SOLAR ECLIPSE

What Is It?
Alignment of Sun, Moon, and Earth

Difficulty Level
Difficult

Description
The number-one astronomical event visible to the naked eye is a total solar eclipse. This experience is so much better than anything else in our list (even seeing the southern lights) that it should have its own category. It is off-the-charts amazing!

Picture this: the Moon slides slowly in front of the Sun and at just the right moment, when you are standing at just the right place, whoosh! A shadow sweeps over you, the sky turns so dark that the stars and planets come out, and the temperature drops 15 degrees in an instant. When you look up at the sky, there is a perfect black circle where the Sun used to be encircled by a beautiful white halo. Wispy tendrils of white gas seem to stream out of the eclipse. This is the corona of the Sun, a $1,000,000°C$ region that suddenly emerges as the Moon blocks the brighter glare of the Sun's disc. If you look closely you may see tiny pink protuberances, little flames of gas called prominences, sticking out like hairs from the eclipsed Sun. The light around you is so utterly weird that you may think the world is ending. Maybe you'll hoot, holler, or even cry. This is totality, and experiencing it will blow your mind.

When viewing any solar eclipse, remember to use proper eye protection. Before and after totality, the Sun will still be so bright that it can damage your eyes. Only viewing during totality is safe.

Getting the Sun and Moon to be in perfect alignment is a rare and beautiful thing. From Earth, the Moon and Sun appear to be almost the same size in the sky. At times, the Moon is just large enough and just close enough to Earth that it can barely block out the entire Sun. The longest a total solar eclipse can last is 7.5 minutes. The precision of totality is so fleeting that most total eclipses only last a few minutes. Then the Moon slides out of alignment, rays of sunlight peek out, and totality is over. You are left in awe, wondering, "When is the next one of these?"

How to Find It

Unlike a comet, meteor storm, or aurora, a total solar eclipse is a predictable, guaranteed "wow" moment. Astronomers can tell you the exact place on Earth to stand and the exact second that it will start. The only thing you have to worry about is clouds, the bane of sky watchers everywhere. The next total solar eclipses in the Southern Hemisphere will be:

- July 2, 2019 (Chile and Argentina)
- December 14, 2020 (Chile and Argentina)
- December 4, 2021 (Antarctica)
- April 20, 2023 (Indonesia)
- July 22, 2028 (Australia and New Zealand)
- November 25, 2030 (South Africa and Australia)

Or you can become an eclipse chaser and travel to the United States in 2024, Iceland in 2026, or Egypt in 2027. Eclipse chasing is a great excuse to visit exotic locations. It is never too early to plan ahead. After you experience your first total solar eclipse, you'll never be the same.

INDEX

About the Author

Author photo by Mary Strubbe

Dean Regas is the astronomer for the Cincinnati Observatory and cohost of the syndicated TV program *Star Gazers*. His writing has appeared in *Astronomy* magazine, *Sky & Telescope* magazine, and *Huffington Post*, and in 2017 he began a podcast called *Looking Up*. Dean is the author of two other books, *Facts from Space!* (2016) and *100 Things to See in the Night Sky* (Northern Hemisphere edition, 2017).